Data Observability
for Data Engineering

Proactive strategies for ensuring data accuracy and
addressing broken data pipelines

Michele Pinto

Sammy El Khammal

BIRMINGHAM—MUMBAI

Data Observability for Data Engineering

Group Product Manager: Reshma Raman
Publishing Product Manager: Heramb Bhavsar
Content Development Editor: Joseph Sunil
Technical Editor: Devanshi Ayare
Copy Editor: Safis Editing
Project Coordinator: Shambhavi Mishra
Proofreader: Safis Editing
Indexer: Subalakshmi Govindhan
Production Designer: Prafulla Nikalje
Marketing Coordinator: Vinishka Kalra

First published: December 2023

Production reference: 1151223

Published by Packt Publishing Ltd.
Grosvenor House
11 St Paul's Square
Birmingham
B3 1RB, UK

ISBN 978-1-80461-602-4
www.packtpub.com

Contributors

About the authors

Michele Pinto is the Chief Technology Officer at Sustainable Brand Platform. With over 15 years of experience, Michele has a keen sense of how data observability and data engineering are closely linked. He started his career as a software engineer and has worked since then in various positions such as Big Data Engineer, Big Data Architect, Head of Data and, in recent years, in the role of Head of Engineering as well as CTO. He strongly believes in hard work and teamwork and loves to create the best possible conditions for his teams to work in an environment that inspires and motivates them to perform at their best.

To Sara, Giulia and Pierino for the time I took from you while working on this book and for the love you give me every day. To my mother Antonia and my father Giuseppe, who have believed in me for more than 40 years.

Thanks to Andy Petrella, Kensu and the Data Owls team for the great challenges and good times we had together, which contributed to the content and soul of this book.

Sammy El Khammal is a Product Manager at Kensu. After studying business across three continents at Solvay, Monash and Waseda, he dove headfirst into the data realm. Over the past five years, he evolved from customer success to Product Management, also making his mark as a public speaker at O'Reilly. His debut book is a testament to this journey, aiming to demystify Data Observability for a broader audience. Beyond his professional pursuits, Sammy is a guitar aficionado and a triathlon aspirant, showcasing a life enriched by varied interests and continuous learning.

I would like to extend my heartfelt thanks to those who have stood by me and offered their unwavering support during this journey. A special acknowledgment goes to my parents, Martine and Abdelmajid, their guidance and love have shaped the person I am today. Additionally, I am immensely grateful to my friends whose companionship and encouragement have been invaluable. Their belief in me and my work has been a constant source of motivation.

About the reviewers

Andy Rabone is an experienced Data Engineering manager, having worked in the fields of data warehousing, data engineering and business intelligence across healthcare, legal, and financial services for the past 15 years. He is currently engineering lead for a critical regulatory transformation project for a large FTSE 100 company, and passionately promotes the value of data observability and data quality in his work. He lives in South Wales with his wife and son, and enjoys indulging his son's early fascination with robotics, following US sports, and playing guitar every now and then.

Sunil Mandowara is a Lead Data Analytics Architect who is helping organizations achieve their data analytics and AI vision by building scalable, observable, and secure data analytics platforms with his 20 years of hands-on experience carrying out 15+ complex data platform implementations in various domains using a variety of technologies, including distributed data processing, big data, REST, data lakes, data security, cloud, SAAS, and microservices.

Across industries, he has set Data teams from the ground up. He has also implemented the data governance and quality, which significantly boosted the productivity of the team and data value.

I want to thank my twins, Manan and Mahi, for allowing me to spend time reviewing this excellent book.

Table of Contents

Part 1: Introduction to Data Observability

1

2

Part 2: Implementing Data Observability

3

4

5

Part 3: How to adopt Data Observability in your organization

6

7

Optimizing Data Pipelines   139

8

Organizing Data Teams and Measuring the Success of
Data Observability   151

Part 4: Appendix

9

Data Observability Checklist   173

10

Preface

Welcome to the world of Data Observability, an essential domain in the landscape of data engineering. Data observability is the art and science of effectively monitoring, managing, and optimizing data pipelines within modern data ecosystems. This discipline is crucial for ensuring pipeline accuracy, reliability, and usability, particularly as businesses increasingly depend on data-driven decision-making.

Our book, Data Observability for Data Engineering, is a specialized resource, focusing on the practical aspects of implementing and maintaining data observability in real-world scenarios. Unlike many existing resources that broadly cover data management and quality, this book delves into the intricacies and best practices specific to data observability, an often under-explored yet essential facet of data engineering.

The inspiration behind this book stems from our extensive experience in the field. We have navigated the evolving realm of data observability, gaining insights and expertise that we are eager to share. This book is a culmination of our journey.

The chapters of this book are carefully structured to guide you from foundational concepts to advanced techniques and strategies. We delve into the core elements of data observability, covering key topics like data quality monitoring, incident management, and the role of automation in data observability. Through real-world examples and case studies, we demonstrate how these concepts are applied in practice and deliver value, ensuring that readers can translate theory into actionable strategies in their professional roles.

As the complexity of data ecosystems continues to expand, the role of data observability becomes increasingly significant in maintaining the integrity and reliability of these systems. This book is intended to be a resource that not only educates but also empowers you to implement and champion data observability within your organization.

We are excited to share our knowledge and experiences with you through this book. We believe that the insights and techniques presented will be invaluable in your journey towards mastering data observability.

Who this book is for

This book is an essential resource for a broad range of professionals engaged in data transformation and preparation.

It is particularly tailored for Data Engineers, Data Architects, Data Analysts, Data Scientists, and Data Product Owners who have encountered challenges with broken data pipelines or unreliable dashboards and are eager to learn best practices for implementing data observability in their systems.

Furthermore, it's an invaluable guide for managerial roles such as Heads of Data, Data Platform Managers, and Data Engineering Managers who are accountable for data quality and process optimization. These leaders, responsible for ensuring data quality and streamlining processes, will find valuable methodologies and insights to enhance consumer trust and improve producer engagement within data pipelines.

Whether you're directly handling data or managing data-focused teams, this book provides essential knowledge and tools to advance data observability in your organization.

What this book covers

Chapter 1, Fundamentals of Data Quality Monitoring, covers a general introduction to data quality and explains the key metrics used to measure it. It will also explain how data quality can be converted to Service Level Agreements (or contracts) to establish trust among data pipeline stakeholders.

Chapter 2, Fundamentals of Data Observability, will complete the user's knowledge of data quality by adding the observability dimension, taking quality to the next level, and explaining how we can improve data quality monitoring to have real-time contextual information on data pipelines.

Chapter 3, Data Observability Techniques, covers how a data engineer can retrieve information from applications at run time. It will be an overview of the existing techniques and will explain their advantages and disadvantages regarding the efficient implementation of Data Observability.

Chapter 4, Data Observability Elements, provides an overview of the elements needed to collect contextual and real-time information from a pipeline. This will cover a description of those elements and showcase an example of how you can collect them within a Python script doing data manipulation.

Chapter 5, Defining Rules on Indicators, introduces the concepts of continuous validation of the data. The reader will understand how rules can be implemented by the data engineer, manually or in the code, to test the data and where such validation rules can be implemented.

Chapter 6, Root Cause Analysis, focuses on the data issues and how adopting the Data Observability approach simplifies and may even automate anomaly detection and troubleshooting. It will provide a method for Data Incident Management and anomaly detection examples.

Chapter 7, Optimizing Data Pipelines, explains how data observability can be used to manage several aspects of the data pipeline lifecycle such as the cost containment in data pipeline maintenance as well as to aim key aspects like automating documentation, managing catalog, mitigating anomalies, and reduce the change risk.

Chapter 8, Organizing Data Teams and Measuring the Success of Data Observability, focuses on how to introduce Data Observability in your team, describing the different kinds of Data Teams, the different types of organizations where these teams must fit, and how to measure the success of this initiative.

Chapter 9, Data Observability Checklist, suggests a method in the form of a checklist to implement Data Observability in the company pipelines, reviewing the common pitfalls and concerns we encountered when implementing data observability in various companies.

Chapter 10, Pathway to Data Observability, closes the book by providing data engineers with a technical roadmap to implement data observability in a first project and then at scale across the organization.

To get the most out of this book

If you want to follow the code example of this book, you will need to have a proper Python environment, completed with Jupyter Notebook installation. However, the book is structured in a way that ensures comprehension of data observability concepts, regardless of hands-on engagement with the code. These examples are supplemental, enhancing understanding for those who prefer a practical approach.

Software/hardware covered in the book	Operating system requirements
Python >= 3.8	Windows, macOS, or Linux
Jupyter Notebook	

If you are using the digital version of this book, we advise you to type the code yourself or access the code from the book's GitHub repository (a link is available in the next section). Doing so will help you avoid any potential errors related to the copying and pasting of code.

Download the example code files

You can download the example code files for this book from GitHub at `https://github.com/PacktPublishing/Data-Observability-for-Data-Engineering`. If there's an update to the code, it will be updated in the GitHub repository.

We also have other code bundles from our rich catalog of books and videos available at `https://github.com/PacktPublishing/`. Check them out!

Conventions used

There are a number of text conventions used throughout this book.

`Code in text`: Indicates code words in text, database table names, folder names, filenames, file extensions, pathnames, dummy URLs, user input, and Twitter handles. Here is an example: " User `ID202` performed a query to create the `customers.marketing` table."

A block of code is set as follows:

```
{"execution_timestamp":"2022-05-03;17:04:09", "user":{"ID202"},
query:"INSERT INTO `customers.marketing` SELECT NAME, EMAIL, AGE,
IS_LOYAL, TOTAL_BASKET FROM `customers.info` INNER JOIN `customers.
orders` ON customers.info.id = customers.orders.id"}
```

When we wish to draw your attention to a particular part of a code block, the relevant lines or items are set in bold:

```
{"NAME":str, "EMAIL":str, "AGE":int, "IS_LOYAL": bool, "TOTAL_
BASKET":double}
```

Bold: Indicates a new term, an important word, or words that you see onscreen. For instance, words in menus or dialog boxes appear in **bold**. Here is an example: ". For instance, **page_visited** has to contain only integer numbers."

> **Tips or important notes**
> Appear like this.

Get in touch

Feedback from our readers is always welcome.

General feedback: If you have questions about any aspect of this book, email us at `customercare@packtpub.com` and mention the book title in the subject of your message.

Errata: Although we have taken every care to ensure the accuracy of our content, mistakes do happen. If you have found a mistake in this book, we would be grateful if you would report this to us. Please visit `www.packtpub.com/support/errata` and fill in the form.

Piracy: If you come across any illegal copies of our works in any form on the internet, we would be grateful if you would provide us with the location address or website name. Please contact us at `copyright@packt.com` with a link to the material.

If you are interested in becoming an author: If there is a topic that you have expertise in and you are interested in either writing or contributing to a book, please visit `authors.packtpub.com`.

Share Your Thoughts

Once you've read *Data Observability for Data Engineering*, we'd love to hear your thoughts! Scan the QR code below to go straight to the Amazon review page for this book and share your feedback.

https://packt.link/r/1-804-61602-8

Your review is important to us and the tech community and will help us make sure we're delivering excellent quality content.

Download a free PDF copy of this book

Thanks for purchasing this book!

Do you like to read on the go but are unable to carry your print books everywhere? Is your eBook purchase not compatible with the device of your choice?

Don't worry, now with every Packt book you get a DRM-free PDF version of that book at no cost.

Read anywhere, any place, on any device. Search, copy, and paste code from your favorite technical books directly into your application.

The perks don't stop there, you can get exclusive access to discounts, newsletters, and great free content in your inbox daily

Follow these simple steps to get the benefits:

1. Scan the QR code or visit the link below:

https://packt.link/free-ebook/9781804616024

2. Submit your proof of purchase.
3. That's it! We'll send your free PDF and other benefits to your email directly.

Part 1: Introduction to Data Observability

In this section, we introduce data quality fundamentals, including key metrics and their application in Service Level Agreements to build trust in data pipelines. We then explore data observability, enhancing data quality monitoring with real-time insights for more effective management of data systems.

This part has the following chapters:

- *Chapter 1, Fundamentals of Data Quality Monitoring*
- *Chapter 2, Fundamentals of Data Observability*

1
Fundamentals of Data Quality Monitoring

Welcome to the exciting world of *Data Observability for Data Engineering*!

As you open the pages of this book, you will embark on a journey that will immerse you in data observability. The knowledge within this book is designed to equip you, as a data engineer, data architect, data product owner, or data engineering manager, with the skills and tools necessary to implement best practices in your data pipelines.

In this book, you will learn how data observability can help you build trust in your organization. Observability provides insights directly from within the process, offering a fresh approach to monitoring. It's a method for determining whether the pipeline is functioning properly, especially in terms of adhering to its data quality standards.

Let's get real for a moment. In our world, where we're swimming in data, it's easy to feel like we're drowning. Data observability isn't just some fancy term – it's your life raft. Without it, you're flying blind, making decisions based on guesswork. Who wants to be in that hot seat when data disasters strike? Not you.

This book isn't just another item on your reading list; it's the missing piece in your data puzzle. It's about giving you the superpower to spot the small issues in your data before they turn into full-blown catastrophes. Think about the cost, not just in dollars, but in sleepless nights and lost trust, when data incidents occur. Scary, right?

But here's the kicker: data observability isn't just about avoiding nightmares; it's about building a foundation of trust. When your data's in check, your team can make bold, confident decisions without that nagging doubt. That's priceless.

Data observability is not just a buzzword – we are deeply convinced it is the backbone of any resilient, efficient, and reliable data pipeline. This book will take you on a comprehensive exploration of the core principles of data observability, the techniques you can use to develop an observability approach, the challenges faced when implementing it, and the best practices being employed by industry leaders. This book will be your compass in the vast universe of data observability by providing you with various examples that allow you to bridge the gap between theory and practice.

The knowledge in this book is organized into four essential parts. In part one, we will lay the foundation by introducing the fundamentals of data quality monitoring and how data observability takes it to the next level. This crucial groundwork will ensure you understand the core concepts and will set the stage for the next topics.

In part two, we will move on to the practical aspects of implementing data observability. You will dive into various techniques and elements of observability and learn how to define rules on indicators. This part will provide you with the skills to apply data observability in your projects.

The third part will focus on adopting data observability at scale in your organization. You will discover the main benefits of data observability by learning how to conduct root cause analysis, how to optimize pipelines, and how to foster a culture change within your team. This part is essential to ensure the successful implementation of a data observability program.

Finally, the fourth part will contain additional resources focused on data engineering, such as a data observability checklist and a technical roadmap to implement it, leaving you with strong takeaways so that you can stand on your own two feet.

Let's start with a hypothetical scenario. You are a data engineer, coming back from your holidays and ready to start the quarter. You have a lot of new projects for the year. However, the second you reach your desktop, Lucy from the marketing team calls out to you: *"The marketing report of last month is totally wrong – please fix it ASAP. I need to update my presentation!"*

This is annoying; all the work that's been scheduled for the day is delayed, and you need to check the numbers. You open your Tableau dashboard and start a Zoom meeting with the marketing team. The first task of the day: understand what she meant by *wrong*. Indeed, the turnover seems odd. It's time for you to have a look at the SQL database feeding the dashboard. Again, you see the same issue. This is strange and will require even more investigation.

After hours of manual and tedious checks, contacting three different teams and sending 12 emails, you finally found the culprit: an ingestion script, feeding the company's master database, was modified to express the turnover in thousands of dollars instead of units. Because the data team didn't know that the metric would be used by the marketing team, the information did not pass and the pipeline was fed with the wrong data.

It's not the first time this has happened. Hours of productivity are ruined by firefighting data issues. It's decided – you need to implement a new strategy to avoid this.

Observability is intimately correlated with the notions of data quality. The latter is often defined as a way of measuring data indicators. Data quality is one thing, but monitoring it is something else! Through this chapter, we will explore the principles of data quality and understand how those can guide you on the data observability journey and how the information bias between stakeholders is key to understanding the need for data quality and observability in the data pipeline.

Data quality comes from the need to ensure correct and sustainable data pipelines. We will look at the different stakeholders of a data pipeline and describe why they need data quality. We will also define data quality through several concepts, which will lead to you understanding how a common base can be created between stakeholders.

By the end of this chapter, you will understand how data quality can be monitored and turned into metrics, preparing the ground for data observability.

In this chapter, we'll cover the following topics:

- Learning about the maturity path of data in companies
- Identifying information bias in data
- Exploring the seven dimensions of data quality
- Turning data quality into SLAs
- Indicators of data quality
- Alerting on data quality issues

Learning about the maturity path of data in companies

The relationship between companies and data started a long time ago, at least from the end of the 1980s, with the first large diffusion of computers in offices. Since computers and data became more and more widespread in subsequent years, the usage of data in companies has gone through a very long period, of at least two decades, during which investments in data have grown, but this was done linearly. We cannot speak of a *data winter*, but we can consider it as a long wait for the spring that led to the explosion of data investments that we have experienced since the second half of the 2000s. This period was interrupted by at least three fundamental factors:

- The collapse of the cost of the resources necessary to historicize and process data (memories and CPUs)
- The advent of IoT devices, the widespread access to the internet, and the subsequent tsunami of available data
- The diffusion and accessibility of relatively simple and advanced technologies dedicated to processing large amounts of data, such as Spark, Delta Lake, NoSQL databases, Hive, and Kafka

When these three fundamental pillars became accessible, the most attentive companies embarked on a complex path in the world of data, a maturity path that is still ongoing today, with several phases, each with its challenges and problems:

Figure 1.1 – An example of the data maturity path

Each company started this path differently, but usually, the first problem to solve was managing the continuously growing availability of data coming from increasingly popular applications, such as websites for e-commerce, social platforms, or the gaming industry, as well as apps for mobile devices. The solution to these problems has been to invest in small teams of software engineers who have experimented with the use and integration of big data technologies and platforms, among which there's Hadoop with its main components, HDFS, MapReduce, and YARN, which are responsible for historicizing enormous volumes of data, processing them, and managing the resources, respectively, all in a distributed system. The more recent advanced technologies, such as Spark, Flink, Kafka, NoSQL, and Parquet, provided a further boost to this process. These software engineers were unaware that they were the first generation of a new role that is now one of the most popular and in-demand roles in software engineering – the data engineer.

These *primal* teams have often been seen as research and development teams and the expectations of them have grown with increasing investments. So, the next step was to ask how these teams could express their potential. Consequently, the step after that was to invest in an analytics team that could work alongside or as a consumer of the data engineers' team. The natural way to start extracting value from data was with the adoption of advanced analytics and the introduction of techniques and solutions based on machine learning. Then, companies began to acquire a corporate culture and appreciate the great potential and competitiveness that data could provide. Whether they realized it or not, they were becoming data-driven companies, or at least data-informed; in the meanwhile, data began to be taken seriously – as a real asset, a critical component, and not just a mysterious box from which to take some insight only when strictly necessary.

The first results and the constant growth of the available data triggered a real race that has pushed companies to invest more and more in personnel and data technologies. This has led to the proliferation of new roles (data product manager, data architect, machine learning engineer, and so on) and the explosion of data experts in the company, which led to new and unexplored organizational problems. The centralized data team model revealed all its limits in terms of scalability and the lack of readiness to support the real problems of the business. Therefore, the process of decentralizing these data experts has begun and, in addition to solving these issues, has introduced new challenges, such as the need to adopt data governance processes and methodologies. Consequently, with this decentralization and having the data more and more central in the company, paired with the need to increase the skills of data quality, what was only important yesterday is becoming more and more of a priority today: to govern and monitor the quality of data.

The spread of teams and data in companies has led to an increase in the data culture in companies. Interactions between decentralized actors are increasingly entrusted via contracts that various teams make between them. Data is no longer seen as an unreliable dark object to rely on if necessary. Each team works daily with data, and data is now a real product that must comply with quality standards that are on par with any other product generated in the company. The quality of data is of extreme importance; it is no longer one problem of many, it *is* the problem.

In this section, we learned about the data maturity path that many companies are facing and understood the reasons that are pushing companies to invest more and more in data quality.

In the next section, we will understand how to identify information bias in data, introduce the roles of data producers and data consumers, and cover the expectations and responsibilities of these two actors toward data quality.

Identifying information bias in data

Let's talk about a sneaky problem in the world of data: information bias. This bias arises from a misalignment between data producers and consumers. When the expectations and understanding of data quality are not in sync, information bias manifests, distorting the data's reliability and integrity. This section will unpack the concept of information bias in the context of data quality, exploring how discrepancies in producers' and consumers' perspectives can skew the data landscape. By delving into the roles, relationships, and responsibilities of these key stakeholders, we'll shed light on the technical intricacies that underpin a successful data-driven ecosystem.

Data is a primary asset of a company's intelligence. It allows companies to get insights, drive projects, and generate value. At the genesis of all data-driven projects, there is a business need:

- Creating a sales report to evaluate top-performing employees
- Evaluating the churn of young customers to optimize marketing efforts
- Forecasting tire sales to avoid overstocking

These projects rely on a data pipeline, a succession of applications that manipulate raw data to create the final output, often in the form of a report:

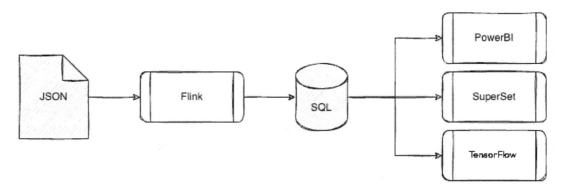

Figure 1.2 – Example of a data pipeline

In each case, the produced data serves the interests of a consumer, which can be, among others, a manager, an analyst, or a decision-maker. In a data pipeline, the applications or processes, such as Flink or Power BI in *Figure 1.2*, consume and produce data sources, such as JSON files or SQL databases.

There are several stakeholders in a pipeline at each step or application: the producers on one hand and the consumers on the other hand. Let's look at these stakeholders in detail.

Data producers

The producer creates the data and makes it available to other stakeholders. By definition, a producer is not the final user of the data. It can be a data engineering team serving the data science team, an analyst serving the board of managers, or a cross-functional team that produces data products available for the organization. In our pipeline, for instance, an engineer coding the Spark ingestion job is a producer.

As a data producer, you are responsible for the content you serve, and you are concerned about maintaining the right level of service for your consumers. Data producers also need to create more projects to fulfill a maximum amount of needs coming from various teams, so producers need to deal with maintaining quality for existing projects and delivering new projects.

As a data producer, you have to maintain a high level of service. This can be achieved by doing the following:

- **Defining clear data quality targets**: Understand what is required to maintain high quality, and communicate those standards to all the data source stakeholders

- **Ensuring those targets are met thanks to a robust validation process**: Put the quality targets into practice and verify the quality of the data, from extraction to transformation and delivery

- **Keeping accurate and up-to-date data documentation**: Document how the process modified the data with instruments such as data lineage and metrics
- **Collaborating with the data consumers**: Ensure you set the right quality standards so that you can correctly maintain them and adapt to evolving needs

We emphasize that collaboration with consumers is key to fulfilling the producer's responsibilities. Let's see the other end of the value chain: data consumers.

Data consumers

The consumer uses the data created by one or several producers. It can be a user, a team, or another application. They may or may not be the final user of the data, or may just be the producer of another dataset. This means that a consumer can become a producer and vice versa. Here are some examples of consumers:

- **Data or business analysts**: They use data produced by the producers to extract insights that will support business decisions
- **Business managers**: They use the data to make (strategic) decisions or follow indicators
- **Other producers**: The consumer is a new intermediary in the data value chain who uses produced data to create new datasets

A consumer needs correct data, and this is where data quality enters the picture. As a consumer, you are dependent on the job done by the producers, especially because your inputs, whether they need to feed another application or a business report, depend directly on the outputs of the producers. Let's look at the different interactions among the stakeholders.

The relationship between producers and consumers

Both producers and consumers are interdependent. Consumers need the raw materials from the producers as inputs and can create inputs for other producers.

Besides this, a producer can have several dependent consumers. For instance, the provider of a data lake will create data that will be the backbone of multiple projects in the data science team.

Conversely, a consumer can use various producers' data to create their data product. Take the example of a churn model, a machine learning project that aims to identify the customers who are about to leave the contract. Those models will use the **Customer Relationship Management** (**CRM**) data from the company, but will also rely on external sources such as the **International Monetary Fund** (**IMF**) to extract the GDP per capita.

In a data pipeline, these two roles are alternating and a consumer can easily become a producer (see *Figure 1.3*). In well-structured data-driven companies, it is often the case where a team will be responsible for collecting data, another team will ingest the data in the master data, and a data analyst team will use it to create reports. The following figure depicts a data pipeline where consumers and producers are the stakeholders:

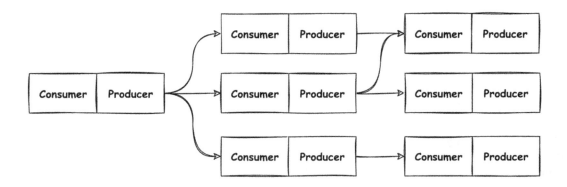

Figure 1.3 – A view of a pipeline as a succession of producers and consumers

As you can see, a pipeline can become complex in terms of responsibilities when several stakeholders are involved.

For a consumer, the quality of the data is key. It is about getting the right tools at the right time, such as in an assembly line. If you work in a car manufacturing company, you won't expect to receive flat tires when it's your turn to work on the car (or maybe it is worse if the chain is late or completely stopped). You can, of course, control the quality of the tires on your own, when you receive them from the tire manufacturer, before putting them on the car. Nevertheless, at this stage, the issue is detected too late as it will eventually slow down car production.

In these interconnected pipelines, issues may arise once the quality of the data doesn't meet the consumers' expectations. It can be even worse if those issues are detected too late, once the decision has already been taken, leading to disastrous business impacts. As a result, the trust of the consumers in the whole data pipeline is eroded, and the data producer becomes more hesitant to deploy the new data application to production, significantly lowering the time to market.

In a large-scale company, even small quality issues at the beginning of the pipeline can have bad consequences on the outcome. Without a good data quality process, teams lose days and months firefighting issues. Finding the cause of an issue or even detecting the issue itself can be painful. The consumer may detect the issue and come back to the producer, asking them to fix the pipeline as soon as possible – not to say immediately. However, without good-quality processes, you may spend days analyzing complex data pipelines and asking for permission to read data from other teams.

Asymmetric information among stakeholders

While the goal of each stakeholder is clear – producers want to send the highest quality data to consumers, and consumers want the best quality standard for their data – who is responsible for the data quality is not clear. Consumers expect data to be *of good quality*, and that this quality is ensured and backed up by the producers. On the contrary, producers expect their consumers to validate and control the quality of data they deserve. This results in a misalignment of objectives and responsibilities.

This is at the root of what we describe as information bias between producers and consumers. They both have *asymmetric information* about data quality. This is a situation where one party has more or better information than the other, which can lead to an imbalance of power or an unfair advantage. The producer wants to deliver quality defined by the customer but needs to receive defined and accurate expectations from them.

The consumer knows the important metrics they want to follow. However, it requires good communication within data teams to ensure these parameters are understood by the producers.

There is also a shared responsibility paradigm: while the data producers bear the responsibility for ensuring quality, the consumers play an important role in providing feedback and setting clear expectations. This shared responsibility is also key to fostering a good data quality culture inside the organization.

Data quality is paramount because it's the cornerstone of trust in any data-driven decision-making process. Just like a house needs a solid foundation to stand, decisions need reliable data to be sound. When data quality is compromised, everything built on top of it is at risk.

With that, we have defined why data quality is important and how it can enforce relationships in a company. Now, we'll learn what data quality is by exploring the seven dimensions of data quality.

Exploring the seven dimensions of data quality

Now that you understand why data quality is important, let's learn how to measure the quality of datasets.

A lot has already been explained about data quality in the literature – for instance, we can cite the DAMA (Data Management Body of Knowledge) book. In this chapter, we have decided to focus on seven dimensions: accuracy, completeness, consistency, conformity, integrity, timeliness, and uniqueness:

Figure 1.4 – Dimensions of data quality

In the following sections, we will cover these dimensions in detail and explain the potential business impacts of poor-quality data in those dimensions.

Accuracy

The accuracy of data defines its ability to reflect the actual data. In a CRM, the contract type of a customer should be correctly associated with the right customer ID. Otherwise, marketing action could be wrongly targeted. For instance, in *Figure 1.5*, where the dataset was wrongly copied into the second one, you can see that the email addresses were mixed up:

date	order_id	email	page_visited	duration	total_basket	has_confirmed
01/01/2021	1aacfa9c-770d	baslie4t@bravesites.com	3.0	700.82	299.1	1
02/01/2021	5d0f80cf-2a32	mhextfz@nba.com	12.0	550.71	617.24	1
03/01/2021	8d2b7e99-dcd9	pgeorgeonax@umich.edu	7.0	195.28	558.07	0
04/01/2021	2f518f9c-39e2	sstannas7d@mtv.com	13.0	353.41	48.65	1
05/01/2021	cd24bd4c-4d46	wprestney1t@google.com.br	11.0	92.82	289.82	1
06/01/2021	3f9be911-5bb7	parmisteadg0@jimdo.com	5.0	503.79	123.86	0

date	order_id	email	page_visited	duration	total_basket	has_confirmed
01/01/2021	1aacfa9c-770d	parmisteadg0@jimdo.com	3.0	700.82	299.1	1
02/01/2021	5d0f80cf-2a32	baslie4t@bravesites.com	12.0	550.71	617.24	1
03/01/2021	8d2b7e99-dcd9	mhextfz@nba.com	7.0	195.28	558.07	0
04/01/2021	2f518f9c-39e2	pgeorgeonax@umich.edu	13.0	353.41	48.65	1
05/01/2021	cd24bd4c-4d46	sstannas7d@mtv.com	11.0	92.82	289.82	1
06/01/2021	3f9be911-5bb7	wprestney1t@google.com.br	5.0	503.79	123.86	0

Figure 1.5 – Example of inaccurate data

Completeness

Data is considered complete if it contains all the data needed for the consumers. It is about getting the right data for the right process. A dataset can be incomplete if it does not contain an expected column, or if missing values are present. However, if the removed column is not used by your process, you can evaluate the dataset as complete for your needs. In *Figure 1.6*, the **page_visited** column is missing, while other columns are missing values. This is very annoying for the marketing team, who are in charge of sending emails, as they cannot contact all their customers:

date	order_id	email	duration	total_basket	has_confirmed
01/01/2021	1aacfa9c-770d	baslie4t@bravesites.com	700.82	299.1	1
02/01/2021	5d0f80cf-2a32	mhextfz@nba.com	550.71	617.24	
03/01/2021	8d2b7e99-dcd9	pgeorgeonax@umich.edu		558.07	0
04/01/2021	2f518f9c-39e2		353.41		
05/01/2021	cd24bd4c-4d46		92.82	289.82	1
06/01/2021	3f9be911-5bb7	parmisteadg0@jimdo.com	503.79	123.86	0

Figure 1.6 – Example of incomplete data

The preceding case is a clear example of where data producers' incentives can be different from customers'. Maybe the producer left empty cells to increase sales conversions as filling in the email address may create friction. However, for a consumer using the data for an email campaign, this field is crucial.

Consistency

The way data is represented should be consistent across the system. If you record the addresses of customers and need the ZIP code for the model, the records have to be consistent, meaning that you will not record the city for some customers and the ZIP code for others. At a technical level, the presentation of data can be inconsistent too. Look at *Figure 1.7* – the **has_confirmed** column recorded Booleans as numbers for the first few rows and then used strings.

In this example, we can suppose the data source is a file, where fields can be easily changed. In a **relational database management system (RDMS)**, this issue can be avoided as the data type cannot be changed:

date	order_id	email	page_visited	duration	total_basket	has_confirmed
01/01/2021	1aacfa9c-770d	baslie4t@bravesites.com	3.0	700.82	299.1	1
02/01/2021	5d0f80cf-2a32	mhextfz@nba.com	12.0	550.71	617.24	1
03/01/2021	8d2b7e99-dcd9	pgeorgeonax@umich.edu	7.0	195.28	558.07	0
04/01/2021	2f518f9c-39e2	sstannas7d@mtv.com	13.0	353.41	48.65	True
05/01/2021	cd24bd4c-4d46	wprestney1t@google.com.br	11.0	92.82	289.82	True
06/01/2021	3f9be911-5bb7	parmisteadg0@jimdo.com	5.0	503.79	123.86	False

Figure 1.7 – Example of inconsistent data

Conformity

Data should be collected in the right format. For instance, **page_visited** can only contain integer numbers, **order_id** should be a string of characters, and in another dataset, a ZIP code can be a combination of letters and numbers. In *Figure 1.8*, you would expect to see @ in the email address, but you can only see a username:

date	order_id	email	page_visited	duration	total_basket	has_confirmed
01/01/2021	1aacfa9c-770d	baslie4t	3.0	700.82	299.1	1
02/01/2021	5d0f80cf-2a32	mhextfz	12.0	550.71	617.24	1
03/01/2021	8d2b7e99-dcd9	pgeorgeonax	7.0	195.28	558.07	0
04/01/2021	2f518f9c-39e2	sstannas7d	13.0	353.41	48.65	1
05/01/2021	cd24bd4c-4d46	wprestney1t	11.0	92.82	289.82	1
06/01/2021	3f9be911-5bb7	parmisteadg0	5.0	503.79	123.86	0

Figure 1.8 – Example of improper data

Integrity

Data transformation should ensure that data items keep the same relationship. Integrity means that data is connected correctly and you don't have standalone data – for instance, an address not connected to any customer name. Integrity ensures the data in the dataset can be traced and connected to other data.

In *Figure 1.9*, an engineer has extracted the **duration** column, which is useless without **order_id**:

duration
700.82
550.71
195.28
353.41
92.82
503.79

Figure 1.9 – Example of an integrity issue

Timeliness

Time is also an important dimension of data quality. If you run a weekly report, you want to be sure that the data is up to date and that the process runs at the correct time. For instance, if a weekly sales report is created, you expect to receive a report on last week's sales. If you receive outdated data, because the database was not updated with the new week's data, you may see that the total of this week's report is the same as last week's, which will lead to wrong assumptions.

Time-sensitive data, if not delivered on time, can lead to inaccurate insights, misinformed decisions, and, ultimately, monetary losses.

Uniqueness

To ensure there is no ambiguous data, the uniqueness dimension is used. A data record should not be duplicated and should not contain overlaps. In *Figure 1.10*, you can see that the same order ID was used for two distinct orders, and an order was recorded twice, assuming that there is no primary key defined in the dataset. This kind of data discrepancy can lead to various issues, such as incorrect order tracking, inaccurate inventory management, and potentially negative customer experiences:

date	order_id	email	page_visited	duration	total_basket	has_confirmed
01/01/2021	1aacfa9c-770d	baslie4t@bravesites.com	3.0	700.82	299.1	1
02/01/2021	1aacfa9c-770d	mhextfz@nba.com	12.0	550.71	617.24	1
03/01/2021	8d2b7e99-dcd9	pgeorgeonax@umich.edu	7.0	195.28	558.07	0
04/01/2021	2f518f9c-39e2	sstannas7d@mtv.com	13.0	353.41	48.65	1
05/01/2021	cd24bd4c-4d46	wprestney1t@google.com.br	11.0	92.82	289.82	1
05/01/2021	cd24bd4c-4d46	wprestney1t@google.com.br	11.0	92.82	289.82	1

Figure 1.10 – Example of duplicated data

Consequences of data quality issues

Data quality issues may have disastrous effects and negative consequences. When the quality standard is not met, the consumer can face various consequences.

Sometimes, a report won't be created because the data quality issue will result in a pipeline issue, leading to delays in the decision-making process. Other times, the issue can be more subtle. For instance, because of a completeness issue, the marketing team could send emails with `"Hello {firstname}"`, destroying their professional appearance to the customers. The result can damage the company's profit or reputation.

Nevertheless, it is important to note that not all data quality issues will lead to a catastrophic outcome. Indeed, the issue only happens if the data item is part of your pipeline. The consumer won't experience issues with data they don't need. However, this means that to ensure the quality of the pipeline, the producer needs to know what is done with the data, and what is considered important for the data consumer. A data source that's important for your project may have no relevance for another project.

This is why in the first stages of the project, a new approach must be implemented. The producer and the consumer have to come together to define the quality expectations of the pipeline. In the next section, we will see how this can be implemented with **service-level agreements** (**SLAs**) and **service-level objectives** (**SLOs**).

Turning data quality into SLAs

Trust is an important element for the consumer. The producer is not always fully aware of the data they are treating. Indeed, the producer can be seen as an executor creating an application on behalf of the consumer. They are not a domain expert and work in a black-box environment: on the one hand, the producer doesn't exactly understand the objectives of their work, while on the other hand, the consumer doesn't have control over the supply chain.

This can involve psychological barriers. The consumer and the producer may have never met. For instance, the head of sales, who is waiting for their weekly sales report, doesn't know the data engineer in charge of the Spark ingestion in the data lake. The level of trust between the different stakeholders can be low from the beginning and the relationship can be poisonous and compromised. This kind of trust issue is difficult to solve, and confidence is dramatically complex to restore. This triggers a negative vortex that can lead to very important effects in the company.

As stated in the *Identifying information bias in the data* section, there is a fundamental asymmetry of information and responsibilities between producers and consumers. To solve this, data quality must be applied as an SLA between the parties. Let's delve into this key component of data quality.

An agreement as a starting point

A SLA serves as a contract between the producer and the consumer, establishing the expected data quality level of a data asset. Concretely, with a SLA, the data producer is committed to offering the level of quality the consumer expects from the data source. The goals of the SLA are manifold:

- Firstly, it ensures *awareness* of the producer on the expected level of quality they must deliver.
- Secondly, the contractual engagement asks the producers to assume their *responsibilities* regarding the quality of the delivered processes.
- Thirdly, it enhances the *trust* of the consumer in the outcomes of the pipeline, easing the burden of cumbersome double checks. The consumer puts their confidence in the contract.

In essence, data quality can be viewed as an SLA between the data producer and the consumer of the data. The SLA is not something you measure by nature, but this will drive the team's ability to define proper objectives, as we will see in the next section.

The incumbent responsibilities of producers

To support those SLAs, the producer must establish SLOs. Those objectives are targets the producer sets to meet the requirements of the SLA.

However, it is important to stress that those quality targets can be different for each contract. Let's imagine a department store conducting several data science projects. The marketing team oversees two data products based on the central data lake of all the store sales. The first one is a machine learning model that forecasts the sales of the latest children's toy, while the second one is a report of all cosmetic products. The number of kids in the household may be an important feature for the first model, while it won't be of any importance for the second report. The SLAs that are linked to the first project are different from the SLAs of the second one. However, the producer is responsible for providing both project teams with a good set of data. This said, the producer has to summarize all agreements to establish SLOs that will fulfill all their commitments.

This leads to the notion of vertical SLAs versus transversal SLOs. The SLA is considered vertical as it involves a one-to-one relationship between the producer and the consumer. In contrast, the SLO is transversal as, for a specific dataset, a SLO will be used to fulfill one or several SLAs. *Figure 1.11* illustrates these principles. The producer is engaged in a contractual relationship with the customer, while the SLO directs the producers' monitoring plan. We can say that a data producer, for a specific dataset, has a unique set of SLOs that fit all the SLAs:

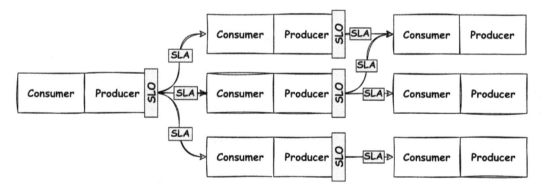

Figure 1.11 – SLAs and SLOs in a data pipeline

The agreement will concern a bilateral relationship between a producer and their consumer. An easy example is as follows: "*To run my churn prediction model, I need all the transactions performed by 18+ people in the last 2 weeks.*" In this configuration, the data producer knows that the data of the CRM must contain all transactions from the last 2 weeks of people aged 18 and over. They can now create the following objectives:

- The data needs to contain transactions from the last 2 weeks
- The data needs to contain transactions of customers who are 18 and over

We'll see how indicators can be set to validate the SLO later.

The SLA is a contract between both parties, whereas the SLO is set by the producer to ensure they can respect all SLAs. This means that an SLO can serve several agreements. If another project requires the transaction data of 18+ customers for 3 weeks, the data team could easily adjust its target and set the limit to 3.

The advantage of thinking of data quality as a service is that, by knowing which dimension of quality your consumers need, you can provide the right adjusted level of service to achieve the objectives. As we'll see later, it will help a lot regarding costs and resource optimization.

Considerations for SLOs and SLAs

To ensure SLAs and SLOs remain effective and relevant, it is important to focus on different topics.

First, as business requirements change over time, it is essential for both parties to regularly review the existing agreements and adjust objectives accordingly. Collaboration and communication are key to ensuring the objectives are always well aligned with consumer expectations.

Documentation and traceability are also important when it comes to agreements and objectives. By documenting those, all parties can maintain transparency and keep track of changes over time. This can be helpful in case of misalignment or misunderstandings.

Also, to encourage producers to respect their agreements, incentives and penalties can be activated, regardless of whether the agreements are met.

What is important and at the core of the SLA/SLO framework is being able to measure the performance against the established SLOs. It is crucial to set up metrics that will assess the data producer's performance.

To be implemented, objectives have to be aligned with indicators. In the next section, we will see which indicators are important for measuring data quality.

Indicators of data quality

Once objectives have been defined, we need a way to assess their validity and follow up on the quality of data on a day-to-day basis.

To do so, indicators are must-haves. Indeed, an SLO will be useless without any indicator. Those indicators can also be called **service-level indicators (SLIs)**. An indicator is a defined measure of data quality. An indicator can be a metric, a state, or a **key performance indicator** (**KPI**). At this stage, data quality is activated and becomes data quality monitoring. The goal of the indicator is to assess the validity of your objectives. It is the producer's responsibility to check whether an indicator behaves well.

Depending on the objective, many indicators can be put in place. If a data application expects JSON as input, the format of the incoming data source becomes an important indicator. We will see techniques and methods for gathering these indicators in *Chapter 3* and learn how a data model can be used to collect those elements in *Chapter 4*.

Data source metadata

We define metadata as data on data. It can be the file location, the format, or even the owner of the dataset. These can be pertinent indicators for executing data applications. If a program expects CSV file format to be triggered but the format has changed upstream to JSON, the objectives can be jeopardized.

Schema

The expected schema of a data source is an important indicator. An application or a pipeline can simply break if the input schema is not correct. The important characteristics of a schema are its field names and field types.

A wrong type, or a deleted column, can lead to missing values in the outcome, or broken applications. Sometimes, the issue is even more sneaky. A Python AI model will only require a feature matrix without column names. If two columns are interchanged, the model will use the wrong value associated with its coefficients and will lead to misleading results.

Lineage

Lineage or data mapping at the application level allows you to express the transformations of data you need in the process. This lineage permits an overview of the data field's usage. Monitoring this lineage can prevent any change in the code base of the application.

Data lineage describes the data flow inside the application. It is the best documentation about what's happening inside each step of a pipeline. This lineage allows you to create a data usage catalog, where you can see who accessed the data, how it was used, and what was created out of it.

Thanks to lineage, you can manage the risk of modifying the application over time as you can easily understand which other applications, data sources, and users will be impacted.

Lineage is also a good indicator if you wish to know if what's happening inside the application can be done or not. Imagine a data pipeline in a bank that grants loans to its customers. The project team may choose to rely on the automatic decision of an AI model. For ethical and GDPR reasons, you do not want any of these decisions to rely on the gender of the customer. However, the feature matrix can be a mix of scores computed from other data sources. Thanks to lineage, you can validate how the feature matrix was created, and avoid any misuse of gender data you have in the data lake.

Application

Information about the application itself is also neglected but an important indicator. The code version (or tool version), as well as the timestamp of execution, can be valuable for detecting data quality issues. An application is a tool, a notebook, a script, or a piece of code that modifies the data. It is fueled by inputs and produces outputs.

An application running a wrong code version may use outdated data, leading to a timeliness issue. This is a perfect example of indicators that will serve a technical objective without a direct link to the agreement. The code version is often out of the scope of consideration for the business team, while it can have a lot of impact on the outcome.

Statistics and KPIs

Broadly speaking, there are many metrics – generic or not – that can become indicators in the context of data quality. Here, we draw a distinction between statistics and KPIs.

Statistics are a list of predefined metrics that you can compute on a dataset. KPIs are custom metrics that are often related to the accuracy of the dataset, which can also relate to a combination of datasets. Let's deep dive into some of the main statistics as indicators:

- **Distribution**: There are many ways to compute the distribution of a dataset feature. For numerical data, the minimum, maximum, mean, median, and other quantiles can be very good indicators. If the machine learning model is very sensitive to the distribution, skewness and kurtosis can also be considered to offer a better view of the shape of the data. For categorical data, the mode and frequency are valuable indicators.

- **Freshness**: The freshness of a dataset is defined by several time-based metrics to define whether the data is *fresh*. This is done by using a combination of the frequency and the timeliness of the dataset:

  - **Frequency**: This metric tells us at what time the dataset was updated or used and allows us to check whether the data was solicited on time. If the dataset is expected to be updated every day at noon and is late, data processes may be wrong if they're launched before the availability of new data rows.

- **Timeliness**: This is a measure of the obsolescence of data. It can be computed by checking the timestamp of the data and the timestamp of the process using it. A reporting process could show no errors while the underlying data is outdated. If you want to be sure that the data you are using is last week's data, you can use a timeliness indicator.

- **Completeness**: Two indicators can help you check whether data is complete or not. The first one is the *volume* of data, while the second one is the *missing values* indicator:

  - **Volume**: This is an indicator of the number of rows being processed in the dataset. The fact that the volume of data drops or sharply increases may express an underlying issue, and can signal an issue upstream or a change in the data collection process. If you expect to target 10 million customers but the query returns only 5,000, you may be facing a volume issue.

  - **Missing values**: This is an indicator of the number or the percentage of null rows or null values in the dataset. A small percentage of missing values can be tolerated by the business team. However, for technical reasons, some machine learning models do not allow any missing values and will then return errors when running.

KPIs are custom metrics that can give an overview of the specific needs of the consumers. They can be business-centric or technical-driven. These KPIs are tailored to work with the expectations of the consumers and to ensure the robustness of the pipeline.

If the objective is to furnish accurate data for the quantity sold, you must set an indicator on the total, at which point you could create a Boolean value that will equal 1 if all the values are positive. This business-driven KPI will guarantee the accuracy of the data item.

A technical KPI can be, for instance, a difference in the number of rows of the inputs and outputs. To ensure the data is complete, the result of a full join operation should ensure that you end up with at least the same number of rows as the input data sources. If the difference is negative, you can suspect an issue in the completeness of the data.

A KPI can also be used to assess other dimensions of data quality. For instance, consistency is a dimension that often requires us to compare several data source items across the enterprise filesystem. If, in a Parquet partition, the **amount** column has three decimals but only one in another partition, a KPI can be computed and provide a score that can be 0 if the format of the data is not consistent.

Examples of SLAs, SLOs, and SLIs

Going back to our previous example, we defined the SLOs as follows:

- The data needs to contain transactions from the last 2 weeks
- The data needs to contain transactions of customers who are 18 and over

To define indicators on those SLOs, we can start by analyzing the dimensions we need to cover. In this example, the SLOs are related to the completeness, timeliness, and relevance of the data. Therefore, this is how you can set up SLIs:

1. Firstly, there is a need to send data promptly. A timeliness indicator can be introduced, making sure you use the data from the two latest weeks at the time of execution. You have to measure the difference between the latest data point and the current date. If the difference is within 14 days, you meet the SLI.

2. Secondly, to make sure the data is complete, you can compute the percentage of missing values of the dataset you create. A threshold can be added for the acceptable percentage of missing values, if any.

3. Thirdly, a KPI can be added to ensure the data contains 18+ consumers. For instance, it can be a volume indicator based on the condition that the age column is greater than 18. If this indicator remains 0, you meet the SLI.

4. Indicators can also be put on the schema to assure the consumer that the transaction amount is well defined and always a float number.

Now that you have defined proper indicators, it is high time you learn how to activate them. This is the essence of monitoring: being alerted about discovered issues. Configuring alerts to notify the relevant stakeholders whenever an SLI fails will help you detect and address issues promptly. Let's see how this can be done with SLIs.

Alerting on data quality issues

Once indicators have been defined, it is important to set up systems to control and assess the quality of the data as per these indicators. An easy way to do so is by testing the quality with rules.

An indicator reflects the state of the system and is a proxy for one or several dimensions of data quality. Rules can be set to create a link between the indicators and the objective(s). For a producer, violating these rules is equivalent to a failed objective. Incorporating an alerting system aims to place the responsibility of detecting data quality issues in the producer's hands.

An indicator can be the fuel of several rules. Indicators can also be used over time to create a rule system involving variations.

Using indicators to create rules

Collecting indicators is the first step toward monitoring. After that, to prevent data quality issues, you need to understand the normality of your data.

Indicators reflect the current state of the dataset. An indicator is a way of measuring data quality but does not assess the quality *per se*. To do so, you need to understand whether the indicator is valid or not.

Moreover, data quality indicators can also be used to prevent further issues and define other objectives linked to the agreement. Using lineage, you can define whether modifying indicators upstream in the data flow can have an impact on the SLA you want to support.

Rules can be established based on one indicator, several indicators, or even the observation of an indicator over time.

Rules using standalone indicators

A single indicator can be the source of detection of a major issue. An easy way to create a rule on an indicator is to set up an acceptable range for the indicator. If the data item has to represent an age, you can set up rules on the distribution indicators of minimum and maximum. You probably don't want the minimum age to be a negative number and you don't want the maximum to be exaggerated (let's say more than 115 years old).

Rules using a combination of indicators

Several indicators can be used in a single rule to support an objective. The missing value indicator in a column can be influenced by the missing values of other datasets. In a CRM, the *Full Name* column can be a combination of the *First Name* and *Surname* columns. If one row of the *First Name* column is empty, there is a high probability that *Full Name* will also fail. In that case, setting a rule on the *First Name* column also ensures that the completeness objective of *Full Name* is fulfilled.

Rules based on a time series of indicators

The variation of an indicator over time should also be taken into consideration. This can be valuable in the completeness dimension, for instance. The volume of data, which is the number of rows you process in the application, can vary over months. This variation can be monitored, alerting the producer if there is a drop of more than 20% in the number of rows.

Rules should be the starting point of alerts. In turn, these alerts can be used to detect but also prevent any issue. When a rule detects an issue, it helps to ensure a trust relationship with the consumer as they will be able to assess the quality of the data before using it.

The data scorecard

With the rules associated with the data source, indicators and rules can be used to create a non-subjective data scorecard. This scorecard is an easy way for the business team to assess the quality of the data comprehensively. The scorecard can help the consumer share their issues with the producers, avoiding the traditional *My data is broken, please fix it!* issue. Instead, the consumer can stress the reason for the failing job – *I've noticed a drop in the quality of my dataset: the percentage of null rows in the Age column exceeds 3%.* It also helps the consumer understand the magnitude of the problem, and the producer to prioritize their work. You won't react the same if the number of null rows is 2% as if the number of missing values has been bumped up by more than 300%.

The primary advantage of a scorecard is that it aims to increase the trustworthiness of the dataset. Even if the score is not the best one today, the user is reassured and knows that an issue will be detected by the producer itself. As a result, the latter gains in reputation. Creating a scorecard for the datasets you produce demonstrates your data maturity. It promotes a culture of continuous improvement within the team and organization.

Also, a scorecard helps in assessing data quality issues. By assigning weights to different dimensions of data quality, the scorecard allows you to prioritize aspects of data quality to ensure the most critical dimensions get the necessary focus.

This scorecard can be created per data usage, which means that a score can be associated with each SLA. We suggest the scorecard is a mirror of the data quality dimensions. Let's look at some techniques for creating such a scorecard.

Creating a scorecard – the naïve way

To start with an easy implementation of the scorecard, you can use a percentage of the number of rules met (rules not being broken) over the total number of rules. This gives you a number between 0 and 100 and tells you how the data source behaved when it was last assessed.

Creating a scorecard – the extensive way

This scorecard is created based on one or several dimensions of data quality. For each dimension and each SLA, you can compute the score of the rules used to test those dimensions. To do so, follow these steps:

1. Identify the data quality dimensions relevant to the data source.
2. Assign weight, or importance, to each quality dimension based on its importance to the business objectives.
3. Define rules for each dimension you want to cover while considering the requirements of the SLAs.
4. Compute a score for each rule. You can use the naïve approach of counting the number of respected rules or you can use a more sophisticated approach.
5. Compute the weighted score by multiplying the score of each rule by the weight of its corresponding dimension and summing all the results.

By visualizing and tracking the created scores, you can easily share them with your stakeholders and compare data sources with each other, as well as detecting trends and patterns.

Let's summarize what we've learned in this chapter.

Summary

In this chapter, we saw why data quality is important. Data quality allows us to prevent and solve issues in data processes. We explored the dimensions of data quality and what measures can be taken.

Next, we analyzed the data maturity path that companies started on years ago and are still taking and how this path is bringing about the urgent need to have an ever-greater focus on data quality.

We also defined producer-consumer information bias, leading to a shift in responsibilities for data pipeline stakeholders. To solve this, we proposed using the service-level method.

First, data quality must be considered as a service-level agreement, which is a contract between the producer and the consumer. These contracts contain the expected level of quality the data users require.

Second, the agreements are processed by the data producers, who will create a set of objectives that aim to support one or several agreements.

Third, to ensure that the objectives are met, the producer must set up indicators to reflect the state of the data.

Finally, the indicators are used to detect quality issues by creating rules that can trigger actions on the side of the data producer through alerts. The validity of those rules can be used to create a scorecard, which will solve the information bias problem by ensuring everyone is well informed about the objectives and the way they are controlled.

In the next chapter, we will see why those indicators are the backbone of data observability and how data quality can be turned into data observability.

2
Fundamentals of Data Observability

Perfect! You have set up new objectives for the marketing report data so Lucy, the marketing analyst we met in *Chapter 1*, is delighted. Your team is now aware of the expectations of the marketing team, and the trust relationship between the marketing and data engineering teams can be rebuilt. You are now thinking about ways your data engineering team will monitor quality, and it will start by analyzing the datasets you provide to the business in order to prove they are correct. Every day, you check the quality of the output table in a Snowflake database. Quickly, an error appears. The values in a column seem to have been divided by a factor of 10. What a shame! Lucy is already using the dataset. Hopefully, you can proactively contact her and avoid errors in her report. Nevertheless, you must now *fix* the issue. This won't be a piece of cake… Looking at the central data catalog, you are already losing hope… How should you start?

Actionable data quality isn't easy. Even if the pipeline is monitored, you can only rely on some spaghetti of lineages and metrics – a lot of linked data sources where, at first glance, nothing seems valuable and useful to you, especially because of the multitude of links. The team is spending a lot of time analyzing the data issue and is even wondering whether it is solely responsible for it, as the database is used for a lot of purposes. You were supposed to implement the panacea to any quality issues. Was data quality a mirage?

Well, your issue is common in a lot of organizations. Once it has been decided to implement a data quality program, it is important to be able to leverage it. To do so, data observability is key.

In the first chapter, we saw the importance of data monitoring to enhance the trust and awareness of a data team. Your consumers can be reassured by adding service level agreements, objectives, and indicators, and you, as the producer, are aware of the needs and usage of the data pipelines you are covering.

Setting objectives is the first step toward data quality. However, the classical way of conducting data quality monitoring comes with some drawbacks. In this chapter, we will start with the main defects of current data quality monitoring methods, and we will learn how data observability makes up for their shortcomings.

In this chapter, we'll cover the following topics:

- From data quality to data observability
- The three principles of data observability
- Data observability in IT observability
- Key components of data observability

Technical requirements

This chapter contains interactive examples that you can also run. All the code examples for this chapter can be found at `https://github.com/PacktPublishing/Data-Observability-for-Data-Engineering`.

From data quality monitoring to data observability

The general way of conducting data quality involves manual and automated checks, also called tests, on process inputs and outputs. In this paradigm, on the one hand, the consumer is responsible for checking the validity of their raw material according to their proper needs – for instance, by validating the schema you are receiving. On the other hand, the producer checks the conformity of the output data regarding consumers' needs by ensuring, for instance, that data manipulation did not deteriorate its completeness. Often, if the data team arranges a well-running data quality program, the inputs won't be checked by the consumers as they expect the inputs to be already validated.

The following figure explains this model; the data quality process ensures that the inputs and outputs are in line with quality expectations:

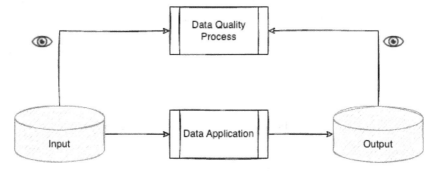

Figure 2.1 – Data quality outside the application

Manual checks are easy when we are handling a single project. They may become more difficult when dealing with a CRM providing data to many data teams or when the data product is part of a bigger pipeline, where the output is needed to feed another process.

In that case, the data quality process must detect the event of any issue as soon as it occurs and may fail its objectives if the issue is detected too late, when the data has already been processed by another producer, or even worse, by the final consumer. If the checks are performed hours or days after the dataset is available to consumers, the error may have already been propagated. This can again result in an erosion of trust. Let's imagine that a sales team report, due to outdated data, led to layoffs in the team. No one would want to feel responsible for this…

Also, once an issue is detected, it is important to know who the issue concerns. As a service level objective can group several service level agreements, it becomes important to keep track of the context of the failure. Moreover, the investigation of the origin of the issue, also called root cause analysis, can be slowed down by noise in the collected metadata if it was poorly gathered. Indeed, collecting indicators from all the datasets must be done wisely in order to avoid having too many or not enough service level indicators.

Finally, manual and automated tests must be performed as often as possible. Often, because of a lack of resources and money, it is decided to perform periodic checks, assuming that if the data is tested and validated once a month, it shouldn't break for a long period. This can have disastrous consequences if we assume that the dataset is correct based on assumptions.

To respond to the need for well-executed data quality monitoring, we introduce the concept of data observability.

The following figure illustrates this concept by putting observability at the application level instead of an external quality process:

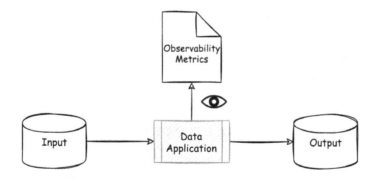

Figure 2.2 – Data observability at the application level

It is time for us to deliver our vision of data observability! And it starts with the word *observable*.

A system is said to be observable when it sends metrics, or observations, to the external world. Thanks to observability, an opaque pipeline can send information about its status to an observer. In the data world, observability allows us to switch from black-box processes to interpretable pipelines and to provide feedback on the status of the process in a timely manner.

Observability gives you information from inside the process itself. It enables a new way of monitoring. It is a manner of knowing whether the pipeline behaves correctly, including by following its own data quality rules.

While data quality focuses on predefining checks on data and making sure those checks are applied to the final datasets, observability focuses on monitoring the data while it is running in the system.

To draw a simple yet insightful analogy, consider the role of a software developer whose primary objective is to engineer software that is devoid of defects. Subsequent to the development phase, the software undergoes a rigorous **quality assurance** (**QA**) process. This crucial step assists developers in identifying and rectifying any bugs before the software is released into production and, ultimately, delivered to the end user. Despite the most thorough QA efforts, it's not uncommon for issues to emerge post-deployment. When such scenarios arise, customers report these issues through various channels. Depending on the severity of the issue, a rapid response mechanism, often referred to as a "hotfix" process, is initiated. This can place significant pressure on the developer to resolve the problem swiftly.

However, a foresighted developer would have preemptively embedded logging mechanisms within the application. These logs act as a diagnostic tool, recording the state of specific variables or the outcomes of certain functions at various points in time. In the event of an issue, these logs prove invaluable; they demystify the application's inner workings, shedding light on the root cause of the problem, thereby facilitating a quicker resolution.

Drawing a parallel to the realm of data observability, similar principles apply. Just as a developer logs critical events in an application, data observability involves meticulously tracking each transformation that data undergoes. This process ensures transparency and accountability, providing a clear audit trail that can be invaluable for troubleshooting and maintaining the integrity of the data life cycle.

In this section, we have seen that a data quality program by itself is not sufficient to solve data issues. To ensure the proper conduct of a pipeline, we have introduced the notion of data observability. This will allow the pipeline itself to inform teams about its behavior. Data observability contains three main principles, so let's explore them in the next section.

Three principles of data observability

By definition, observability comes from the system itself and reporting its state to the outside world. In data observability, the root of the system is the data process. A process is almost always an application, so a script, a notebook, or a program, for instance. The application reports its inner activity, which allows the external observer to understand what happens inside the process. More than creating a data product output, a layer of data observability allows the application to produce data on its proper execution.

Data observability comes with three main principles:

Data observability = Contextual + Synchronous + Continuous validation

This is all about avoiding using unnecessary observations, avoiding data error cascade, and ensuring non-propagation of known issues.

Let's explore these principles in detail:

- **Observability must be put into context**: This point is about contextualizing your indicators. Indeed, a wide range of indicators, or observations, can be collected. However, all those indicators are useless if not put into a certain context. Observability, by providing information from the application itself, gives the first context of the collected information. The following figure shows several pipelines in an enterprise ecosystem:

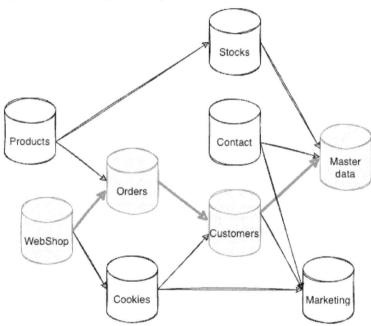

Figure 2.3 – Observability in context allows you to isolate the pipeline from other projects

As in *Figure 2.3*, let's imagine a series of pipelines feeding a master database that's used for many projects. A first pipeline (in red) can send customer order data to the table. Other pipelines add information to the master data. In the final table, we receive data from many subprocesses. Doing data quality on the final data table can mislead the monitoring, as indicators will be collected on columns on which the pipeline has no incidence. Having the context of data transformation will bring many advantages. For a specific project, instead of monitoring all the columns, you would only want to check the columns of interest. Producers can easily understand the background of any data issue and identify the projects where it happened and the projects it impacts.

- **Observability is synchronized with the process**: Secondly, observability, coming from the application, is synchronized with the execution of the application. By doing this, the time needed to detect any issue is considerably reduced. This low-latency strategy avoids the classical error cascade and simplifies the root cause analysis. Indeed, the produced outcomes are always aligned with the metrics and logs collected from the application. Synchronization also means that you will only monitor data that is used in or produced by the application at the time the application is triggered.

Figure 2.4 illustrates the synchronization principle. If, in a pipeline, the quality checks on **Application 1** are performed after **Application 2** has run, the potential errors in **Application 1** will already have been propagated, leading to a waste of resources. Synchronized data observability allows us to monitor the pipeline while running the application and ensure the non-propagation of errors:

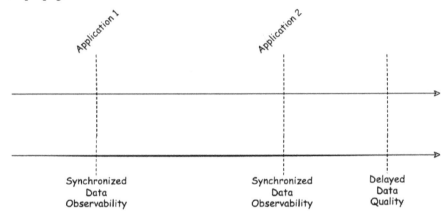

Figure 2.4 – Data observability is synchronized

- **Observability allows continuous validation of the data**: Finally, observability allows us to *continuously validate* the constraints imposed on the data. In fact, the system itself can use its proper metrics to check its behavior. The rules can be applied at runtime, which allows us to immediately validate the inputs or produced datasets. Continuous validation includes observability as a component of the data management life cycle. The data is processed to create outputs, but also to validate them, as you can see in *Figure 2.5*:

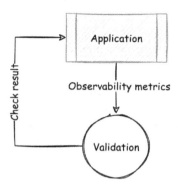

Figure 2.5 – Continuous validation in data observability

The principles of data observability — contextual, synchronous, and continuous validation — can significantly enhance data quality by providing a comprehensive, timely, and dynamic approach to monitoring and validating data processes.

Firstly, contextualizing observability ensures that data is monitored and validated within its relevant operational context. This targeted approach prevents the dilution of efforts and resources on irrelevant data points, thereby improving the focus and effectiveness of QA measures. For instance, monitoring only the relevant columns in a database that a particular process affects reduces noise and improves the precision of quality metrics.

Secondly, synchronizing observability with the data processes ensures that any issues are detected in real time, closely aligning with the application's execution. This immediacy in detection prevents the propagation of errors through the data pipeline, thereby maintaining the integrity of downstream data. It mitigates the risk of error cascades, which can compound in complexity and cost over time.

Lastly, continuous validation embeds quality checks throughout the data life cycle, not just at endpoints. This principle ensures that data is constantly scrutinized against predefined rules and constraints, fostering an environment where data quality is not a one-time checkpoint but an ongoing guarantee. Continuous validation acts as a persistent guardian, ensuring that data meets the necessary standards at every stage of its journey.

In summary, these principles of data observability create a robust framework for maintaining high data quality. They ensure that monitoring is relevant, timely, and integrated into the data life cycle, thereby fostering a proactive and efficient approach to data quality management.

Data observability in IT observability

Observability is not a new asset in the monitoring toolkit. Indeed, making a system observable is already implemented by the operational and DevOps community to report on application or system availability. Observability helps us to monitor complex distributed systems by collecting logs, metrics, or traces.

If we take observability in the broadest sense, every component of the data pipeline can be monitored — not only the data, but the application, the system, or even the users. Data observability helps us to build the bridge between the data engineering world and application and system observability, with some common elements allowing a deeper analysis of data issues in the context of full observability.

Besides the problems that come from the data itself, other issues may arise — for instance, from an unresized cluster, a failing or outdated application, and so on. The combination of all kinds of observability can increase the quality of the investigation.

Data observability by itself may not be sufficient to analyze all the issues that may arise in a data pipeline. Other forms of observability may be needed, and those can have links with data observability too. *Figure 2.6* explains the place of data observability in IT observability:

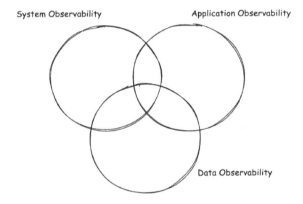

Figure 2.6 – Data observability and other kinds of observability

As part of the general concept of IT observability, we distinguish the following:

- Application observability: Application logs, errors, exceptions, and so on. Knowing about the application can help solve issues.

- System observability, also called infrastructure observability: The size of the cluster, number of available CPUs, and so on. All of this can have an impact on the production of output data by delaying it, for instance.

- Data observability: Containing metadata about the datasets used in a certain context — schema, lineage, metrics, and so on.

In the next section, we will visualize data quality issues in an application.

Key components of data observability

In this section, we will see some examples of data observability metrics that are collected from inside applications and issues that can be raised from such quality issues. We will focus on detecting issues and to do so, we are going to create visuals of data observability issues in a Jupyter notebook.

If you want to follow the example, you can find it in the `Chapter2` section of the GitHub repository. The name of the notebook is `Visualise_Observability_Issues.ipynb`.

In this part, we will focus on a timeliness, a completeness, and an accuracy issue.

The dataset that we provide is a basic example of marketing and sales data. The data represents the orders made on a web shop and consists of the following fields:

- `date`: The date of the order
- `guid`: A unique ID for the order
- `email`: The email address linked to the order
- `page_visited`: The number of pages the customer visited on the website
- `duration`: How long the customer took to prepare the basket of items they'd like to buy and access the payment page
- `total_basket`: The amount of the order
- `has_confirmed`: A Boolean variable indicating whether the user has paid the amount or not

Here is a subset of the dataset:

date	order_id	email	page_visited	duration	total_basket	has_confirmed
2021-01-01	1aacfa9c-770d-4d90-9763-6551c8d89f55	baslie4t@bravesites.com	NaN	700.82	299.10	1
2021-01-01	5d0f80cf-2a32-4d1a-8bc8-e608822e95d4	mhextfz@nba.com	12.0	550.71	617.24	1
2021-01-01	8d2b7e99-dcd9-4883-bf42-061035735a24	pgeorgeonax@umich.edu	7.0	195.28	558.07	1
2021-01-01	2f518f9c-39e2-4d57-8e96-1409f3619e53	sstannas7d@mtv.com	13.0	353.41	48.65	1
2021-01-01	cd24bd4c-4d46-46c8-8259-c31ccfa3b3fa	wprestney1t@google.com.br	11.0	92.82	289.82	1

Figure 2.7 – Subset of the marketing and sales data

This dataset is updated on the fly, and the data is exported every month in order to create reports or to feed the data lake with new raw data for ML projects. Let's imagine you oversee data lake provisioning. This data source of the data lake is your product, and the users and the marketing team are your consumers.

The application simply works like this: every month, a Python script takes a CSV file from a bucket – in this case, the `data` folder – and ingests it into the data lake within the `order` table.

The contract between the application owner and the marketing team

It has been agreed that the produced data will serve several campaigns:

- The first campaign, *Customer conversion*, sends emails to customers who haven't completed their order. The project team expects to retrieve, each month, the list of all order IDs and emails for all orders having `has_confirmed = 1`. The **service level agreement** (**SLA**) for this producer-to-consumer relationship is the following:

  "*We want complete data of all the customers having an incomplete order process for the last month.*"

- The second campaign, a data science project called *Keep buying*, aims to predict whether the customer will confirm the order or not based on their duration on the website, the number of pages they visited, and the amount of the basket. The contract is expressed like this:

  "*We need accurate duration, total_basket, page_visited, and has_confirmed.*"

The first project cannot afford to overlook any row of data, as all the orders of the month need to be processed. The second one is more flexible but will require strong commitment regarding the values of the columns.

These SLAs will be converted into **service level objectives** (**SLOs**). Here are the two metrics you have decided to monitor:

- Data must be complete: It needs to contain `order_id`, `email`, `duration`, `total_basket`, `page_visited`, and `has_confirmed`
- Data must be *fresh*: The data sent to the data lake has to contain the events of the past month

As you can see, these SLOs perfectly match the stricter agreements between the producer and its two consumers.

The pipeline is now in production. In *Part 2: Implementing Data Observability*, we will see how to use different tools and techniques to do so. For this chapter, we will focus on monitoring inside the notebook in order to find known issues. This will show examples of data issues that are sometimes easily detectable and sometimes underlying. These issues will be used to prepare future rules and refine the **service level indicators** (**SLIs**). These will be indicators supporting the objectives — for instance, the components of the schema or the number of missing values.

The following sections will describe failing checks. Let's see which problem can be detected by adding some observability!

Observing a timeliness issue

A timeliness or freshness issue arises when the data has not been updated for a certain period. This can be very problematic when you are dealing with outdated data or incomplete data.

The program is executed for the first time in January. This first run will create a SQL data table that will be filled with future executions.

In this process, an error has occurred. We'll see soon how this error could have been detected and prevented. This error can only be confirmed by a test. Here, we are plotting the mean of the duration within one day. We can observe that there are missing values in the middle of the month, and more specifically, between January 17 and January 23. It seems the data hasn't been updated in this period.

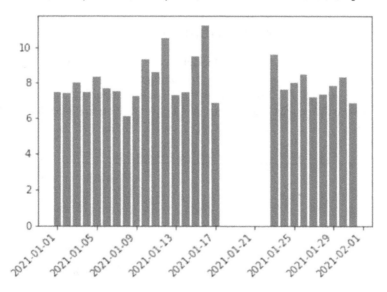

Figure 2.8 – Example of a freshness issue

What can be described as a completeness issue has been discovered through a freshness analysis. Indeed, this issue would have been difficult to detect going by the volume (number of rows) indicator. As it is the first month, no comparison with another execution is possible, and as the number of rows in the input is the same as the output, we cannot assume the program deleted rows. However, by analyzing the data points per day, we can clearly detect a freshness issue, leading us to stamp the January dataset as incomplete.

Observing a completeness issue

Another issue that may arise is a change in the schema. A modified schema can lead to missing keys in a `join` statement, for instance, so that the application simply breaks.

The pipeline is executed for the second time in February. This time, without observability, the error has been quickly detected. Indeed, this schema change issue is caught as soon as the application starts running, as the architecture of the pipeline doesn't allow it to feed the SQL database with any other schema than the one it expects. Knowing that, the check is performed by the code itself and implements a circuit-breaker, breaking the application so that the table cannot be updated.

The exception arises as `OperationalError: table orders has no column named email_customer`.

This said, an additional check could have been performed before writing the DataFrame to the database. In other applications, where data is first manipulated, it makes sense to implement the circuit-breaker as soon as possible to avoid useless and costly data processes.

Observing a change in data distribution

The accuracy of the data can also be responsible for data issues downstream in the pipeline. For the second project, `page_visited` is an important variable that has to be accurate. We will see later how the indicators can be defined. The distribution of the variable is important. Hopefully, you already have two months of data to define what can be considered *genuine*. Plots are incredible tools in order to visualize anomalies:

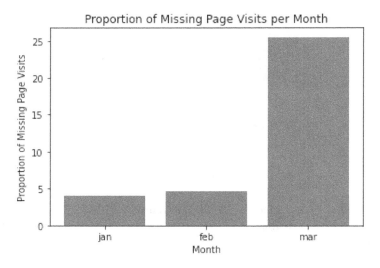

Figure 2.9 – Observing a change in the data distribution

Plotting the percentage of NULL values in the `page_visited` column allows us to see that we have an abnormal number of nulls in March, increasing from under 5% in January and February to more than 25% in March. At this stage, we cannot make any assumptions on the cause of the issue. It could be an error in processes upstream or a change in the customer behavior, for instance, due to new software allowing the web browser to disable cookies.

This anomaly will not affect your treatment, but the SLAs require that you warn the consumer about this issue. As we will see in *Chapter 6*, anomaly detection can be combined with data observability to detect such issues.

As a final word, please note that the issues we have detected are also the base of new contracts with your own producers. As a data engineer, knowing the expectations of your customers is the best way to define your own needs.

Data observability in the enterprise ecosystem

As we have seen in this chapter, through the context, the synchronization with the process, and the continuous validation of data, data observability manages to ground all the potential that often cannot be expressed with traditional data quality approaches. In this section, in particular, we'll understand how businesses can take direct advantage of it.

Investments in data quality are becoming more and more huge, as well as the perception that keeping control of data quality is already a critical aspect of the success of a business. This sense of a need for protection is transforming into investments in qualified personnel and, often, in expensive and sophisticated tools that promise to help companies acquire the means and skills necessary to guarantee certified data products on which to make reliable decisions on an ongoing basis.

If, on the one hand, it is true that enthusiasm and attention are growing exponentially, on the other, these investments have sometimes led to disappointing results:

- As we have already learned, monitoring the quality of data is only the first big step but sometimes it is not conclusive or is inefficient

- Companies often invest in tools that promise interesting solutions but are not accompanied by sufficient and correct knowledge or processes

- Observing and remediating data quality issues is complex, and often, companies and teams still struggle to understand how much this is a determining factor of company competitiveness

- As often happens during the software development cycle, quality is implemented with dedicated teams and dedicated and isolated phases during the software development life cycle

- Many companies have not yet understood what an efficient and scalable data quality process is from an operational and economic point of view

- The data quality ecosystem is still relatively immature, as new technical processes and tools are still appearing on the market

These are just some of the aspects that can also lead to skepticism, so it is important to stop for a moment and understand how data observability can help us to achieve our objectives and to maintain high enthusiasm and hype in the challenge that data quality imposes on us.

Measuring the return on investment – defining the goals

It is often complex to monitor success and especially **return on investment** (**ROI**), particularly in projects that involve different teams and the most disparate business areas, such as a project dedicated to increasing the quality of data in a company. How do you determine the success of a team, the purchase of a tool, or the adoption of a new internal process? It is not easy, but for sure the only answer is to define the objectives and measure them, always.

OK, so let's define the most important high-level goals that drive data observability projects:

- Reducing the number of incidents affecting the quality of data

- Improving the incident response and recovery of data products

Great – now, let's better define why these goals are important and what metrics we can measure to monitor the effectiveness of our efforts to achieve them.

Reducing the number of data incidents

Well, goal number one – the one perhaps anyone would define first – is to try to reduce the number of incidents affecting the quality of data, and the incidents directly impacting our external and internal customers. In the end, if a manager or a team decides to spend time, money, and energy on the quality of their data, it is mainly to avoid incidents that occurred in the past as much as possible. To measure this objective, we can monitor the following metrics:

- **Mean time between failures** (**MTBF**): This is the average time that passes between two distinct incidents. It is particularly useful for measuring the reliability of your data product. It is very easy to calculate: take the entire period of time we are considering (net of downtime periods) and divide it by the number of incidents in the same period.

- **Change failure rate** (**CFR**): This is the probability that a new deployment in production could fail. In other words, it is the number of new failed releases out of the total number of releases. This is a fundamental metric to monitor the ability to deploy and make new changes or implementations to your data pipelines. As the name of the metric suggests, it is the ability to manage change.

- **Deployment frequency**: This is the average time between two deployments. This is not only essential to monitoring how frequent the new deployments are in production but, above all, it is fundamental as a key control indicator for the MTBF and CFR metrics. Indeed, for example, you could discover that the MTBF is very high thanks to the very low number of deployments, therefore, you have a very low deployment frequency.

Great – we are now able to measure how good we are at reducing the number of incidents caused by our data. Are we sure it is enough?

Improving the incident response and recovery of data products

Monitoring and avoiding data incidents is the most important goal, of course. But unfortunately, it is not possible to zero out the likelihood of an incident happening. Therefore, it is equally important to constantly improve your incident management – in other words, the ability to identify and resolve incidents. Let's see what metrics can be used to measure this goal:

- **Mean time to recovery (MTTR)**: This is the average time that your application will take to recover completely from an incident. It is a rather high-level metric that gives us the possibility to measure and monitor the goodness of the process and its ability to resolve an incident from the moment it occurs until its complete resolution. The MTTR is a fundamental metric because, with a single indicator, it is possible to check the efficiency of the resolution of incidents, but it is also a metric that alone does not allow the detailed monitoring of every single phase that the incident management process comprehends.

- **Mean time to detect (MTTD)**: This is the average time between the exact moment a data incident happens to the moment when the incident is detected by a system or a human. This metric allows you to control one of the steps that cannot be analyzed by the MTTR and that is, among other things, one of the most important ones – the capacity to identify the anomalies as quickly as possible.

In this chapter, we learned that, like every investment and project, we must define and monitor metrics from the beginning. Data teams must not only concentrate their efforts on the implementation of their data observability applications and processes, but they must also plan and implement the collection and analysis of data useful for monitoring and sharing metrics that are related to the success of their activities. This is the only way to demonstrate to the rest of the company the effectiveness and goodness of the data observability initiatives, as well as the real and tangible return on investments made by the company.

Summary

In this chapter, we learned about the ties between data quality and observability. We saw that data quality monitoring by itself is often not sufficient to ensure good data and trust in the pipeline.

We introduced the concept of data observability, which will help the monitoring of the data application by applying these three principles:

- Observability is put into context: Data issues must be put into context in order to avoid interfering with other lineages and applications

- Observability needs synchronicity: The sooner you detect the issue, the better you avoid other applications modifying the data source, as long as the appropriate mechanisms to do so are implemented in your pipeline

- Observability allows continuous validation: Use rules to validate data and ensure data quality at runtime

We learned how to measure the success of our projects and investments in data observability, identifying the fundamental objectives and metrics to monitor in order to pursue and demonstrate the effectiveness of our efforts aimed at reducing the number of data incidents and improving the incident response and recovery of our data products. We have also been through a real-life example in order to apply what we have learned so far. Issues such as freshness, schema, and distribution changes were described.

In the next chapter, we will see techniques and methods used to apply the principles of data observability. You will discover how the techniques of data observability differ from pure quality checks.

Part 2: Implementing Data Observability

This section covers advanced topics in data observability. It delves into techniques for data engineers to gather runtime information from applications, comparing their pros and cons for effective observability implementation. Then it will discuss essential elements for collecting contextual, real-time data from pipelines. The end of the section introduces continuous data validation, explaining how data engineers can implement and integrate monitoring rules manually or in code.

This part has the following chapters:

- *Chapter 3, Data Observability Techniques*
- *Chapter 4, Data Observability Elements*
- *Chapter 5, Defining Rules on Indicators*

3
Data Observability Techniques

As we have already learned, sometimes, data quality is still not considered a determining factor during the life cycle of data, or the design and development of new data pipelines.

When a data team has to design and build a new data pipeline, there are many aspects to focus on:

- Data sources – input and output
- Scalability, reliability, and performance
- **Total Cost of Ownership (TCO)**
- Security
- Operation and maintainability
- Compliance with data regulations

But, often, what is missing is the adoption of data quality and observability by design – in other words, defining the specifications regarding what and how to monitor from the beginning of the design of the new data pipeline. These expectations are often not well defined and agreed upon between producers and consumers and even though this might seem unusual, data teams often only develop this sense of need for data quality and observability over time – that is, when the first incident happens in production.

In this chapter, we will look at the different techniques and methods available in our data observability toolset and assess their pros and cons:

- Analyzing the data
- Analyzing the application
- Monkey patching your libraries
- Advanced techniques for data observability – distributed tracing

Analyzing the data

The simplest and most immediate way to monitor what is happening is to focus on monitoring your data. At the beginning of our data quality journey, it seems correct to start with the data, usually because data is exactly what our customers will be using.

But as we saw in the previous chapter, the data itself is often not sufficient to identify and prevent anomalies; for example, even if we are sure that the data we are producing is correct, it is possible that in the meantime, we are ignoring that our new data pipeline is slowly increasing the execution time as well as the usage of hardware resources. Today, we may not notice an anomaly because it is not a critical problem and data is produced as expected, but it could soon become an issue before we realize it. Indeed, what appears to be a pipeline execution issue may soon be converted into a timeliness issue.

On the other hand, our main concern is to be aware of what is happening in our data. So, what does it mean to analyze data to monitor its quality?

Suppose we have a table of customers we would like to monitor. What we could do is list which checks are most useful for monitoring the quality of the data contained in the table:

- The number of rows produced per day
- The total number of rows in the input dataset
- The number of distinct values within the `country` column
- The distinct number of client IDs versus the number of rows
- The maximum and minimum values for the `age` column

For each of these checks, we can write a query to be executed against the appropriate tables, such as our customer table, to retrieve the metrics.

These metrics must then be logged in a dedicated database containing the historical metrics, which are useful for analyzing trends and analyzing any anomalies.

Monitoring data asynchronously

A rather usual approach to monitoring data is to build or buy dedicated tools that do nothing but interrogate and analyze the state of the databases continuously or at specific times of the day.

Let's imagine we build a tool dedicated to monitoring the status of our databases and we call it DataWatch.

The following figure shows the main components contained in a classical asynchronous data monitoring approach:

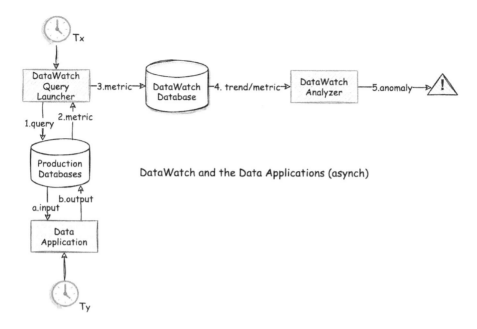

Figure 3.1 – Overview of the asynchronous data analyzer

We decided to schedule DataWatch every day, but now, we have to decide at what time it is convenient to schedule it. The most obvious choice is to execute it right after all the data pipelines that have the greatest impact on the business have been executed and completed. Satisfied with the monitoring provided by DataWatch, after a few days, we decide to schedule it many times a day in such a way as to cover the monitoring process throughout the day.

From this, it is evident that our data applications and our data monitoring tool are decoupled and that the data monitoring process is executed asynchronously from the execution of the data applications.

The following is the flow of our data application, which is a classical **Extract, Transform, Load (ETL)** batch application:

1. It starts every morning at 6 A.M.

2. It extracts data from the input data sources, for instance.

3. It transforms and stores the data in the output data sources.

Now, let's look at the flow of DataWatch, which is scheduled every day at 9 A.M:

1. It triggers several data quality checks as queries against the tables and the databases to be monitored.

2. For each query, it retrieves the metric as a result of the executed query.

3. This metric is stored in dedicated storage that contains all the historical metrics that were gathered from the previous executions.

4. For each query, the metric is analyzed to check whether the expectations are satisfied.

5. If the check fails, an alert is triggered to notify us of the anomaly detection.

Now that we have a clear overview of DataWatch, what it does, why, and how, we have our first tool that is dedicated to helping us monitor our data. Now, it's time to see whether this approach is as good as it seems.

Pros and cons of the asynchronous data analyzer method

Let's consider the pros and cons of this method.

Advantages

There are different pros to analyzing the data asynchronously using an external tool:

-   It is pretty easy: to start, you just need to build a simple script that's used to retrieve, store, and analyze the metrics

-   No rocket science here – everyone in the team can write and instrument some queries, even a non-technical team member

-   It is possible to have a high degree of custom and fine-grained control

Disadvantages

As well as those advantages, there are also big disadvantages:

-   In the end, you are building an additional data application, and like any other data application, it must be monitored, maintained, evolved and, unfortunately, bug-fixed when needed.

-   It's not trivial to build and maintain an application like that. If at the beginning it seems like a silver bullet for your data monitoring problems, after a while, you will quickly discover that you need a lot of time and resources behind it.

-   It's hard to scale. Executing all the checks in the same execution and adding more and more checks becomes time-consuming and also difficult to scale in terms of effort and also in terms of resources needed. To gather some metrics, you have to execute heavy queries against your production database (when possible, use at least one read-only replica dedicated to this purpose).

-   Hidden complexity: If at first glance it seems a simple approach, then it probably hides a considerable amount of complexity. For example, when you need to monitor some custom and complex cases, you may have to write and maintain very sophisticated queries and scripts that are hard to implement and maintain.

- Only analyzing the data could be a short-sighted approach to monitoring data quality as you are missing the entire context. You don't know which applications are performing transformations, both as consumers and producers, on the data sources that are being monitored.

- The data analyzer needs access to all the data sources to monitor. In some companies or business units, it could be a blocker as it may be against internal processes.

- As shown in *Figure 3.2*, the last but perhaps the most important limitation of this approach is the asynchronism between data applications and data monitoring. Often, when an anomaly is identified, it's too late as the data that contains anomalies is already being consumed by other data applications or customers:

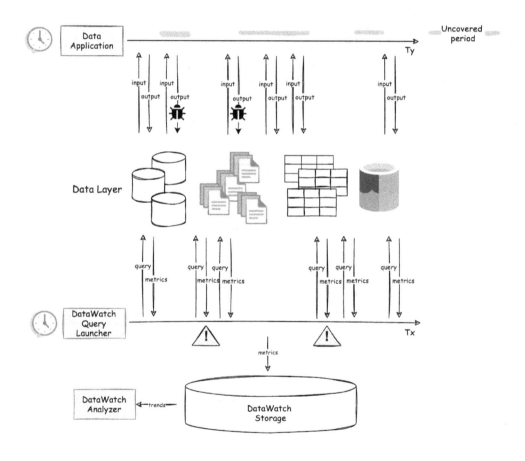

Figure 3.2 – Asynchronism between data applications and data monitoring

Comparing the list of pros and cons, it's easy to understand that in some cases, this asynchronous method is not a suitable approach, even though it's one of the most commonly adopted. If we do not have experience and knowledge of how to monitor data quality, this seems like a good starting point, but the reality is that it hides complexity and several limitations.

Monitoring data synchronously

Asynchronous data monitoring has turned out to be a false friend. Behind an apparent simple implementation, we have discovered and listed several cons and understood how limited and problematic this approach is.

Teams starting with this approach usually start wondering how to iterate and improve the process. Not all problems of asynchronous data analysis can easily be overcome but there are others that, with some smart intuition, can be completely solved or made less problematic.

The first problem to solve is that an asynchronous check cannot guarantee that the input data has been validated. This is because the checks on that source may not have been performed yet. To bypass this, it is possible to synchronize the execution of our application with data control.

As shown in the following figure, the quality check that was performed asynchronously previously is now performed by the application itself in synchrony:

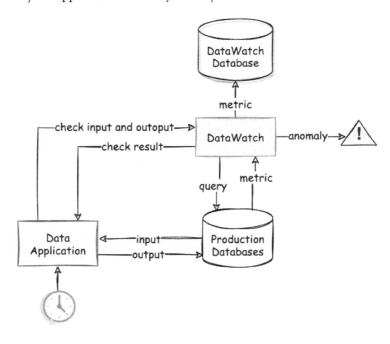

Figure 3.3 – Overview of the synchronous data analyzer

Let's look at the main steps that this approach involves:

1. The application runs and invokes all the data quality checks on the input data sources.

2. If the checks give a negative result, the application can continue and process the data as expected; otherwise, it can decide whether to stop or simply notify the anomaly but continue processing data, especially in the case of non-critical anomalies.

3. At the end of data processing and storing, ETL invokes more data quality checks that are useful for guaranteeing sufficient data quality on the data it has just produced (yes, these are essentially our SLOs – do you remember them?).

4. If a check returns a negative result, the application ends successfully and silently; otherwise, the application will stop with a negative result or report the anomaly.

This approach brings several improvements compared to asynchronous data analysis:

- **Timeliness**: Synchronizing the data applications with data quality monitoring causes the discovery times of the data anomalies to drastically reduce to zero. This entails different advantages, including avoiding the execution (and consequently, the re-execution) of several data pipelines with incorrect data. Consequently, it prevents incorrect data from being delivered to our customers without us realizing it.

- **Scalability**: It is no longer strictly necessary to centralize the execution of the checks in a few daily executions. The checks can be distributed throughout the day.

- **Ownership**: If desired, each application can have its own set of dedicated checks, and like the data applications, each check will have an owner.

- **Maintainability**: Coupling the checks with the data application makes it easier to automatically maintain the data quality checks.

- **Context**: By triggering the checks from the application, it is easier to keep track of the context in which the checks are performed (environment, application, and resources that have been used and are available).

However, several limitations that were already discovered during the asynchronous approach remain uncovered:

- **Complexity**: We still have to implement and maintain complex queries

- **Impact on the performance of the application**: This approach puts the execution of the checks inside the execution of the application but with the risk of extending the execution time of the application

- **Maintenance**: DataWatch, our data analyzer tool, remains a dedicated data application tool to be maintained and evolved

- **Accessibility**: Our data analyzer will still need to access all data sources to be monitored

After examining the diverse advantages of analyzing the data and exploring various approaches, let's now delve into the fundamentals of analyzing an application.

Analyzing the application

A common way to understand what happens in an application is to replay its course after it's run. A good example would be a SQL application. When you query a SQL database, for instance, through a JDBC connector, you are creating access logs in the database. These logs may contain lots of information, especially regarding who has queried the database, what they queried, when it was executed, and sometimes information on how long it took to process the query, how many bytes were retrieved, and so on.

This situation is explained in *Figure 3.4*. Users are continuously querying a central SQL database. This creates a log file, which is a kind of journal that contains the records of the queries:

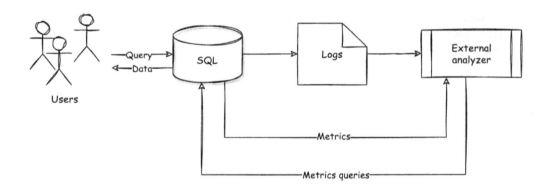

Figure 3.4 – Logging strategy for a SQL logs analyzer

This said, these logs can be extremely valuable for observability purposes. By using strategies to retrieve and analyze the logs, the data team can rebuild data transformation and, by adding some intelligence, construct queries to retrieve stats and metrics.

Not only can the logs of the database be used, but also the logs of the applications themselves. Let's take the example of a dbt Cloud pipeline. dbt is a program that allows you to do the transformation part of ETL. Behind dbt is the creation of a SQL query that can be exposed through a `Manifest. JSON` file. This file, available through an API, will contain all the logs we need to analyze the program.

The *external analyzer* is a program that will process these logs, try to investigate what was performed in the application, and query the database itself for different purposes. Let's see what is important to keep in mind for such an application.

The anatomy of an external analyzer

The external analyzer, as its name suggests, is an external tool that runs asynchronously and aims to process events that are the inputs of the application. The elements of this application are described in *Figure 3.5*:

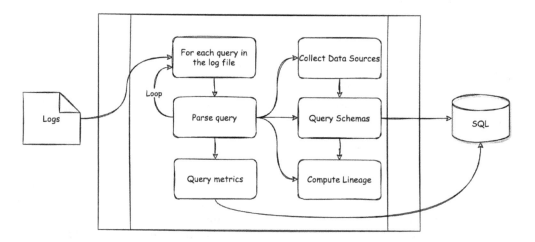

Figure 3.5 – What happens inside the analyzer

The idea is to use the log file and process each line to understand what happened inside. Inside the external analyzer, the following occurs:

1. The data sources are taken out of the query.

2. The database is queried to retrieve the schema.

3. The lineage is computed using the query.

4. A query is created to retrieve metrics or observations from the SQL database.

Let's look at a clear example by considering the following log:

```
{"execution_timestamp":"2022-05-03;17:04:09", "user":{"ID202"},
query:"INSERT INTO `customers.marketing` SELECT NAME, EMAIL, AGE,
IS_LOYAL, TOTAL_BASKET FROM `customers.info` INNER JOIN `customers.
orders` ON customers.info.id = customers.orders.id"}
```

This log contains enough information to reproduce what happened on May 3. User `ID202` performed a query to create the `customers.marketing` table. Let's apply the previously presented method to this log line:

1. First, we'll need to extract all the data sources that were used in the query. Tools such as the `sqlparse` Python library can easily do this. Here, we can see that we have three data sources: the produced data source, which is the `customers.marketing` table, and two consumed data sources, namely `customers.info` and `customers.orders`.

2. Once you know which data sources are used and produced, you can start retrieving the schemas. Hopefully, you can count on tools to do so. Recent SQL databases contain the `INFORMATION_SCHEMA.COLUMNS` view, which contains information such as the column name, type, and nullable policy. If your database doesn't provide you with this metadata, you can also extract sample rows to infer the schema. The schema of the output data source can be expressed like this:

   ```
   {"NAME":str, "EMAIL":str, "AGE":int, "IS_LOYAL": bool, "TOTAL_
   BASKET":double}
   ```

3. The next step is computing the lineage. We'll go through the different types of lineages in the next chapter. Depending on your needs, table-level lineage will be sufficient. It can be done by correctly pinpointing the inputs and outputs among the identified data sources. In our case, the `customers.info` and `customers.orders` tables are the inputs of `customers.marketing`. This lineage is inferred from the query available in the logs.

4. Finally, it's time to compute some metrics and observations from our data. Because we know the schema and the location of the table, we can easily prepare queries. Based on the data type, different metrics can be computed. The following table shows how the query is built based on the schema. In *Chapter 4, Data Observability Elements*, we will look at the interesting indicators such a query can provide:

Field	Type	SQL Query
		SELECT
NAME	STR	sum(case NAME when null then 1 else 0 end) AS name_nullrows,
EMAIL	STR	sum(case EMAIL when null then 1 else 0 end) AS email_nullrows,

Field	Type	SQL Query
AGE	INT	min(AGE) AS age_minimum, avg(AGE) AS age_mean, max(AGE) AS age_maximum,
IS_LOYAL	Bool	count(distinct IS_LOYAL) AS levels_is_loyal,
TOTAL_BASKET	Double	min(TOTAL_BASKET) AS basket_minimum, avg(TOTAL_BASKET) AS basket _mean, max(TOTAL_BASKET) AS basket _maximum,
		FROM `customers.marketing`

Table 3.1 – Examples of SQL queries for statistics

Now that we've seen what the method is, let's dig into its advantages and disadvantages.

Pros and cons of the application analyzer method

The SQL analyzer application is a good method if you wish to understand your data usage. However, you cannot consider this technique as perfect data observability as it misses the synchronous principle. We will see how this method interacts with observability shortly.

Advantages

The analyzer method presented in this section comes with some advantages.

One of the most compelling benefits of the SQL analyzer method is its non-invasive approach to data analysis. By conducting queries outside the main application, it ensures that the core data processes remain untouched. This segregation of analysis from operation is crucial in maintaining the integrity and performance of the primary data processes, making it an ideal choice for environments where uninterrupted data flow is paramount.

The SQL analyzer method exhibits a high degree of compatibility with a diverse array of vendor tools and databases. Its ability to seamlessly integrate with systems that offer access log sharing or provide an API is a significant advantage. This flexibility enables organizations to leverage existing infrastructure and tools, thereby reducing the need for extensive overhauls or the adoption of new, potentially disruptive technologies to enable observability. The SQL analyzer aims to make an opaque system transparent.

Also, if you are not seeking perfect accuracy in the data source metrics, this tool is sufficient for simple data governance as it helps you understand what happens with the data without the need to start a data quality program, such as lineage or data access.

Disadvantages

First, data observability must be performed at runtime. By using an external application, you may collect the wrong data because of the asynchronous run of the external collector. Since data tables may be continuously updated, you cannot be sure that the situation surrounding the table at the time you are computing metrics is the same as when the data table was queried in the original application. In the period between the query's execution and the analyzer being run (even if it's short), a lot of other queries may have interfered with the data table. Because of that, the wrong metrics may be computed, and you cannot assume them to be the source of truth.

Secondly, the context of the application may be lost by analyzing all the queries one after the other. Without contextual information, troubleshooting errors can be ineffective.

Thirdly, this tool, by design, cannot be used to perform continuous validation. One reason is that it cannot be executed at runtime, leading to a lag between the query's execution and the data validation test. Also, because the chance of false positives is too high, validation cannot be trusted so far from the execution time.

Last but not least, this technique involves creating and maintaining a new application with its own development cycle, costs, and privileges. A whole project team has to be dedicated to the subject. Also, the application needs access to the concerned databases, meaning that access rights have to be granted, slowing down the development process in rigid large-scale companies.

To conclude this topic, we can say that, for a data observability use case, the advantages of the tool are not sufficient to outweigh its drawbacks. However, fixing some drawbacks by collecting metrics as close to the execution time as possible can help mitigate the risks.

In the following section, we will present a method that respects the three principles of data observability as much as possible.

Monkey patching your libraries

Monkey patching is a method that involves wrapping the functions or methods used in a program and augmenting them with some capabilities. This method for data observability relies on decorating the code with additional capabilities that will log what's happening inside the application at runtime.

Principles of monkey patching for data observability

Let's take a Python program. In this script, a series of instructions is written using the `pandas` library. These steps include reading a CSV file, renaming a column, and dropping another column before finally saving the DataFrame to a new file. For each step, the program calls a function or a DataFrame method. This is what you can see in *Figure 3.6* if you look at the original program.

In the monkey patching technique, a call to each function must be recorded to trace and analyze the events. It's like taking note of each event that modifies the data and consolidates them at the end. If you are using pandas, all functions and methods should be augmented so that you can keep a record of what's happening inside, and you have to do so for all transformations on the DataFrame:

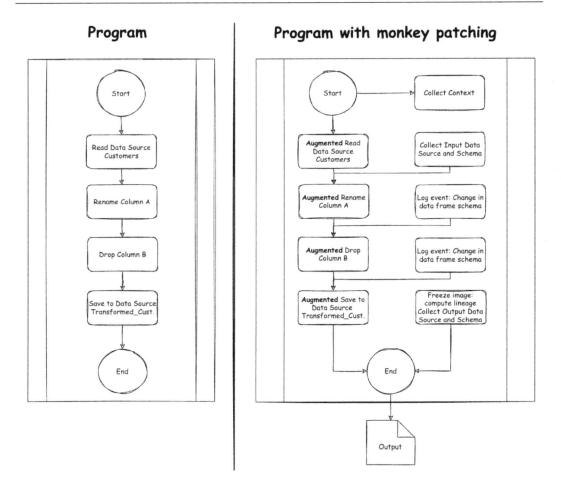

Figure 3.6 – Monkey patching in a script

Next, we'll see what changes have to be added at the microscopic and macroscopic levels. Let's start with the microscopic level, which is the brick that creates a program.

Wrapping the function

To monkey patch the script, each step has to be augmented. Augmenting or wrapping a function involves modifying it to return a different behavior than what's been foreseen. Let's do this with a famous pandas function: `read_csv`. Let's have a look at the following piece of code:

```
import logging
def read_csv(*args,**kwargs):
    import pandas as local_pd
    original_result = local_pd.read_csv(*args,**kwargs)

    data_source_location = args[0] or kwargs['filepath_or_buffer']
    data_source_schema = original_result.describe()
    data_source_format = 'CSV'

    print(f'''The {data_source_format} Data Source
    located in {data_source_location},
    with schema {data_source_schema} is used as an input ''')

    return DataFrame.using(original_result)
```

In this code, we have created a function called `read_csv` based on pandas' eponymous feature. Our `read_csv` function will augment the original one.

The first step involves computing the original result of the function by applying the pandas reader. At this stage, `original_result` is the same object that you would have created by using the `pandas.read_csv` function.

After creating this object, you can start adding logging capabilities to collect the following, for example:

- The file location, by gathering the correct `args` or `kwargs` arguments.
- The schema, by using the pandas method to describe the DataFrame. This method can also be used to retrieve statistics about the DataFrame. The methods is performed on `original_result`.
- The format, which is CSV, which we can derive from the name of the function.

This wrapped function will augment the pandas function each time you call it, allowing you to collect observations.

The result that is returned is not the original result, but an augmented version. In this case, the `DataFrame` object is one where all the methods have been wrapped to add logging capabilities.

By doing this for each function or method, and for all the steps of the program, you will be able to constitute the macroscopic view. Let's take a closer look.

Consolidating the findings

Once every function has produced its logs, it's time to consolidate them. Then, they have to be saved. This can be done by writing them to a file or a server every time a new data source is created or updated. In our pandas example, the logs will be consolidated when the developer writes a file using, for instance, to_csv. To keep track of observability, we propose the following steps:

1. Define the metrics you want to follow – for instance, the data source and schema. A broader list of elements worth gathering will be provided in *Chapter 4, Data Observability Elements*.

2. At the start of the script, you can gather the context of the application, such as its version, the environment in which it is running, and so on. This is the first step in the program when using monkey patching, as shown in *Figure 3.6*. At this stage, tracking is enabled inside a client.

3. While reading a data source, the augmented read function will take care of logging the data source you read. If it uses the read_csv() augmented pandas function, you can infer which data source is used by analyzing the location path you have as an argument of the function. It is also easy to retrieve its format, as well as the schema, as we saw in the previous section. The client collects and saves the logs for later.

4. Each time a function modifies the DataFrame, the client is updated with the information. For instance, if a column is modified, renamed, or dropped, it keeps track of it.

5. Once a data source has been written or updated, the logs are consolidated in event order so that we can create the lineage and save it to an external source.

Building a tool for collecting, aggregating, and visualizing those logs will enable observability by allowing us to create a historical register of all executions.

Pros and cons of the monkey patching method

This method comes with a big advantage compared to the others: it fully respects the three data observability principles. However, it also comes with some issues.

Advantages

As stated earlier, the main advantage of monkey patching is that it perfectly respects the data observability principles.

First, by overriding the original functions and methods, this technique uses the script environment to be executed, collecting data at runtime with no delay. It is, among the techniques that we have presented, the one that introduces the least latency between the data execution and the observability treatment.

Second, by being nested inside the data program itself, this technique allows you to collect comprehensible and exhaustive context, allowing you to easily identify the framework within which the data issue happens.

Third, monkey patching the function allows you to enable continuous validation by adding circuit breakers to the code. We talked about adding logs to the function, but wrappers can also be included to check data SLOs and return an error in case the SLI indicates that one of the rules was violated. This can prevent the program from starting a costly execution if the result of it is going to fail.

Also, once the framework has been developed, the library can be used for all the projects using it. As this technique is agnostic, it doesn't depend on the project itself: the new framework or library builds a set of methods that will help include observability in your data programs. Once the libraries have been wrapped, it becomes easy to implement them in any piece of code, as we will see later in this book.

Last but not least, monkey patching does not require any external application and thus requires no extra privilege to access the data sources. The metrics are computed with the same program as the one used for its execution.

Disadvantages

The main issue is that this technique is time-consuming to develop as an effective monkey patch should wrap all the functions manipulating the data. A lack of support in the function or method can result in important information being skipped and consolidation being impossible. However, depending on the observability metrics you want to gather, this problem can be overcome by combining monkey patching and lines of code.

Another issue comes from the fact that the analyses of the data and the computation of the metrics are done synchronously with the execution of the job. This may introduce some latency, depending on the size and volume of data.

Finally, this technique is used for tools and programs that rely on programming languages. Monkey patching requires that you have access to the necessary source code – for instance, a library – to develop your framework.

We can say that the benefits of this method make it the most convenient for including data observability in data pipelines.

However, monkey patching is not always possible, especially for distributed systems where a client cannot be used in all instances. Some other methods exist to make up for this. In the next section, we will describe one of them.

Advanced techniques for data observability – distributed tracing

As we have seen, there are many ways to implement data observability. It is interesting to dedicate some space to analyze how it is possible to track and collect metrics and metadata within a distributed tracing system.

In the last decade, the adoption of architectures with microservices has become more and more common. Developers and DevOps teams have understood that the adoption of complex and monolithic systems is difficult and expensive to maintain as well as not being scalable. However, the adoption of microservices brings some new challenges, such as the ability to monitor and debug transactions distributed between different services.

To comply with the need to observe what happens between distributed transactions, the distributed tracing mechanism was created for monitoring and profiling distributed applications. This is especially useful for debugging anomalies such as errors or performance issues. **Distributed tracing** is a method dedicated to tracing the entire path of a request from its source, passing through all the applications involved and ending at its destination.

OpenTracing, in turn, is an initiative that was born to support developers and DevOps in implementing distributed tracing without necessarily having vendor lock-in. OpenTracing is not a standard mechanism to describe the behavior of our systems as many believe but it is a vendor-agnostic API that developers and DevOps can reuse to instrument a mechanism inside their applications that works out of the box to perform distributed tracing. It contains the following:

- API specifications for tracing
- Several language-specific libraries
- Vendor-specific wrappers

We can say that OpenTracing is a collection of tools you can use to instrument, generate, collect, and export telemetry data (metrics, logs, and traces) to help you analyze your software's performance and behavior.

To better understand how OpenTracing works, let's imagine that we have an HTTP POST request that triggers several operations and sub-operations:

1. Database read
2. Database write
3. Publish message:

    A. Consume message:

        i.   Database read
        ii. Database write:

(trace)

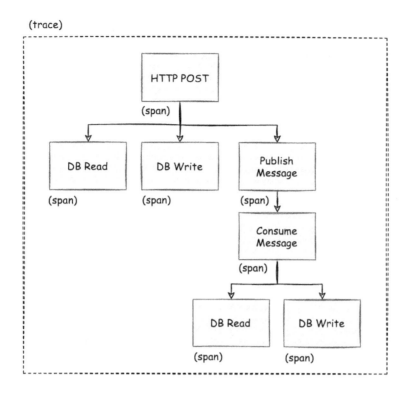

Figure 3.7 – Example of a trace – an HTTP POST request and the related operations

If the entire system is instrumented with OpenTracing, we'll be able to see and monitor the entire operation (that is, the trace) and all the individual connected operations (that is, the spans). Considering the temporal variable, it's easier to visualize traces and the relative spans with a time axis. When you instrument your applications with OpenTracing, you have to send this information to a distributed tracing system that will provide an easy way to gather this timing data. Features include collecting and looking up this data:

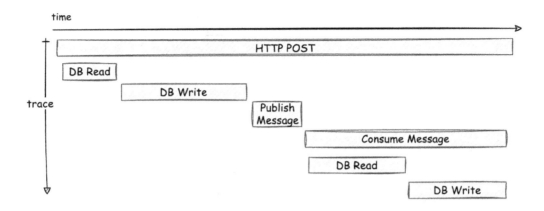

Figure 3.8 – Our HTTP POST trace visualized with a time axis

The key point is that if we can instrument the entire application, then we can also retrieve and track the metrics regarding the quality of the data that our applications are reading and producing.

The idea is to enrich the OpenTracing spans with information related to our data quality. To do so, there are different approaches, the most intuitive one, as represented in the following diagram, is to write a custom wrapper of the OpenTracing API that can gather and track these metrics:

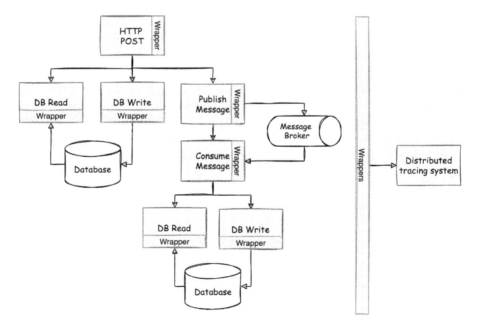

Figure 3.9 – Data observability with distributed tracing

For example, if our **Java Virtual Machine (JVM)** application is reading data from a SQL server, we can implement a wrapper on top of the already existing OpenTracing instrumentation for JDBC. It will be able to do the following:

- Connect to and read data from the SQL server
- Generate the span related to the JDBC query
- Calculate the metrics about the data (for example, the min and max values for each column read) and put them in the span

Advantages

This approach has its pros and cons. The main advantages are as follows:

- You can link distributed tracing with data observability; it's a very powerful weapon
- You can easily do data observability on each pipeline that is already instrumented with distributed tracing
- It works out of the box for several technologies and languages
- By adopting OpenTracing, you can take advantage of a large open source community

Disadvantages

The cons of this approach are as follows:

- You still have to implement a wrapper to enrich the spans with your metrics and, in doing so, you must take care of the different versions of the technologies you are integrating into
- You have to instrument OpenTracing in your applications, which can affect the performance of the application
- At the time of writing, OpenTracing doesn't provide a standard mechanism for tracing information about data

Since OpenTracing was designed to trace the entire path of a request from its source, with the main purpose of monitoring services, it fits perfectly with the fundamentals of data observability. In the future, a natural evolution of OpenTracing could be for it to support data observability while also leveraging the huge community behind this interesting project.

Summary

In this chapter, we looked at different ways of implementing data observability in our data pipelines by weighing the pros and cons of each method.

The first method we considered relies on analyzing the data source, be it synchronously or asynchronously, with the application's execution modifying the data. While this approach seems easy to implement, maintaining an external application can be an issue, and the observability metrics that are gathered may be misleading as they do not obey the three data observability principles – contextual, synchronous, and continuous validation.

A variant of this technique is to start from what the application can tell you about its data transformation by analyzing logs produced during the execution and replaying the run to collect metrics. Again, this method is not fully in line with data observability and introduces complexity outside the original application.

The method that covers the most data observability principles involves the application telling the observer about what's happening with the data it uses and produces in real time: monkey patching. We provided an example of this technique, which logs all the data transformations of a program and summarizes them as part of the process.

Finally, we described more complex methods, such as OpenTracing, which allows us to implement data observability in a distributed system.

In the next chapter, we will focus on which elements, or observability metrics, we should consider collecting when enabling data observability in a pipeline.

4

Data Observability Elements

In the previous chapter, we covered the methods that can be used to collect observability metrics in the context of a data application. We will now focus on the observations themselves. What do you need to collect to keep the data application under control?

In the general observability paradigm, which involves collecting data, the application, and the application's infrastructure, as described in *Chapter 2, Fundamentals of Data Observability*, we saw that observability metrics can be gathered from diverse sources. In *Chapter 3, Data Observability Techniques*, we learned how to extract information directly from data applications. In this chapter, we will focus on which metrics can be collected from the data application itself. We will list and describe all the elements that can be used as **service-level indicators** (**SLIs**) of the data. We will learn how to add SLIs in *Chapter 4*.

Using an open source library, based on the monkey patching methods presented in *Chapter 3*, we will create our first log file containing observations of the data inside the application. This log file will be the basis of any further analyses a tool could do based on data observability.

By the end of this chapter, you will have a deep knowledge of the elements that are needed so that you can claim that your application is observed. You will have also applied monkey patching techniques in Python.

We will cover the following topics:

- Prerequisites and installation requirements
- Static and dynamic elements
- Defining the data observability context
- Getting the data source metadata
- Mastering lineage
- Computing data observability metrics
- Data observability for AI models

Technical requirements

To be able to run the example provided in this chapter, you will need a Python environment with the `pip` repository manager installed. The example has been tested with Python 3.7+. You can find this book's GitHub repository at `https://github.com/PacktPublishing/Data-Observability-for-Data-Engineering`.

Prerequisites and installation requirements

This chapter introduces many concepts that are used in the `kensu-py` open source library. If you want to follow how the log file was generated, please refer to the notebook in this book's GitHub repository, in the `chapter 4` section.

If you are familiar with Python and want to run the example by yourself, we advise you to create a virtual environment; see `https://docs.python.org/3/library/venv.html` for more details.

To install the necessary libraries for this chapter, run `pip install -r requirements.txt` in this book's repository's directory.

Kensu – a data observability framework

In this chapter, we will be working with an open source library called Kensu. It is a Python library that interacts with data transformation libraries to generate observations and send them to the Kensu platform.

Kensu allows you to collect and process data observations through the Kensu platform. You can try out the platform by logging into the trial edition, which comes with a free plan for a limited period. To learn more, go to `https://kensu.io`. You do not have to use it for this chapter as we will only have a look at the agent.

Kensu comes with various agents and collectors. An agent is a piece of code that is nested in your development environment and generates runtime observations. In this chapter, we'll focus on the Python agent, which we'll use to explain how the Kensu model is created.

kensu-py – an overview of the monkey patching technique

`kensu-py` uses monkey patching to modify the data manipulation functions of Python libraries and log some events on the fly.

In the first notebook, `Compute_YTD`, we will use the `orders` dataset we described and introduced in *Chapter 2, Fundamental of Data Observability*, and manipulate data with `pandas` to create a report and save it, as shown in the following code sample:

```
import pandas as pd
df_jan = pd.read_csv("../data/jan/orders.csv"), \
        parse_dates=['date'])
df_feb = pd.read_csv("../data/feb/orders.csv"), \
        parse_dates=['date'])
df_feb = df_feb.rename({'email_customer':'email'},axis=1)
data_YTD = pd.concat([df_jan,df_feb])
data_YTD.to_csv("../data/feb/ordersYTD.csv",index=False)
```

This script reads the `orders.csv` data from January and February to concatenate it and save it in a **Year-To-Date (YTD)** file called `ordersYTD.csv`. Also, the February dataset must be modified as a column does not match the January dataset. The `email_customer` column must be renamed `email`.

The monkey patching technique will keep every event (data transformation) in this script in memory to create observability elements. We will see how it does this throughout this chapter.

To install `kensu-py`, you need to execute `pip install kensu` in your Python environment. The examples provided in this book are based on version 2.7.1 of the library, which you can obtain by running `pip install kensu==2.7.1`.

The `kensu-py` source code is available on GitHub at `https://github.com/kensuio-oss/kensu-py`.

`kensu-py` overwrites other Python functions. Monkey patching can be included inside data manipulation libraries. The structure of `kensu-py` is pretty simple – the library is organized into modules that group the functions while following the major data manipulation libraries. This means that, for each library, the corresponding Kensu module has to be used, as follows:

```
import pandas as pd --> import kensu.pandas as pd
```

Importing Kensu-tuned libraries instead of vanilla libraries allows us to monkey-patch the functions and methods with minimal code changes.

In the notebook, you can view the augmented code base by running the following command:

```
from kensu.utils.kensu_provider import KensuProvider
K = KensuProvider().initKensu()
```

Let's see which elements we can collect from this type of program.

Static and dynamic elements

First, let's focus on what we consider as a data observability element in this case. A data observability element is a piece of data you can retrieve from the running application that aims to make the pipeline observable. If it can be monitored, the same element can then become a SLI.

It's important to make a clear distinction between two categories of observations: static and dynamic.

The set of static elements represents the assets, whereas the set of dynamic elements represents the usages of those assets. For instance, the application will be ranged in the static category, while the application will be run in the dynamic one.

The static elements correspond to all the observations that can be manually reported by a human documenting their data usage because they represent assets that are located and (virtually) accessible and can used or reused. Dynamic observations are often linked to the execution or usage of static elements and are intimately linked to the notion of runtime, the period when the application interacts with other static resources.

The static elements of data observability that we'll classify and define in this chapter are as follows:

- The data source and its schema
- The application, its code base, version, and author
- The lineage
- The project, environment, and user

The dynamic elements we'll look at are as follows:

- The execution of the application
- The execution of the lineage
- The metrics collected during the application's execution

The dynamic elements will create links between the different static elements. These elements are included in a data model that's used by kensu-py, based on primary keys that link the elements with each other.

These elements are linked to each other through a data observation model. The model we propose here is the one that will be used by Kensu and will serve as a common model for many enterprise systems.

The data model we are going to describe is shown in the following diagram:

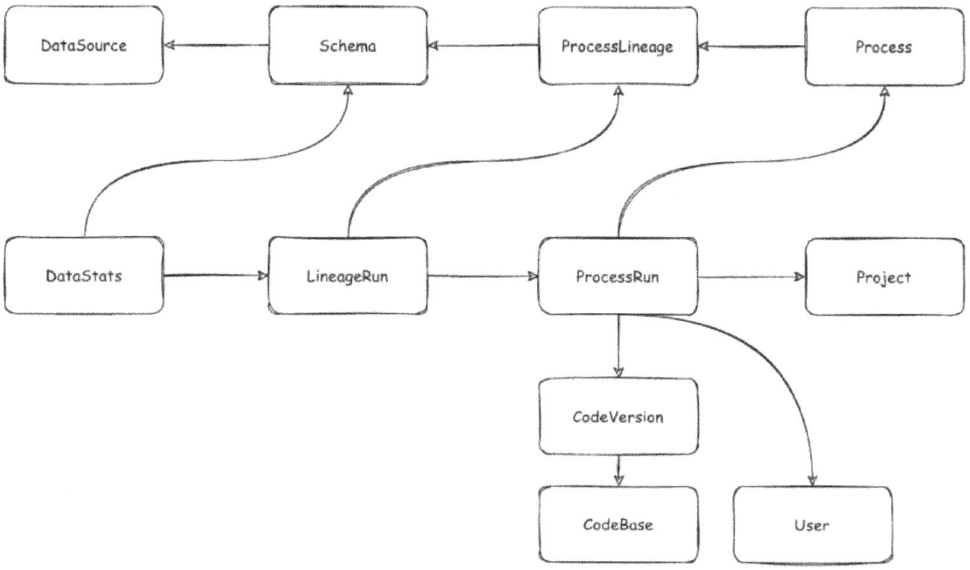

Figure 4.1 – Data model for observability elements

Let's explore these elements one by one.

Defining the data observability context

Following the data observability principles, the context of data manipulation is important. Now is a good time to define what we mean by context in data observability. We can define the context as the set of circumstances of the data transformations – in other words, they are the metadata that can help you understand how and where the data transformation or manipulation happened. The context will tell you which application manipulated the data, when it was manipulated, who executed the manipulation, what triggered it, and so on. This context should give you all the necessary pieces of information while you're debugging the code or the data issue, both upstream (root cause analysis) and downstream (impact analysis).

Long story short, the context is the background of the application. It starts at the beginning of the script or program execution and lasts until all the data transformations the application was supposed to perform are completed.

The execution context is composed of several elements. Let's analyze them and see how they are handled in the open source Kensu library.

Application or process

The application, also called a process, is the core of the execution context. The application is simply a name that corresponds to the script, notebook, job, or program.

Here are some examples of applications:

- A program such as Informatica, Cognos, or Excel
- A sheet name, such as in Excel or Tableau
- A script for a Python or Scala script
- A Jupyter or Databricks notebook

In our example, we will define the name of the notebook as the application. The application is a static element. It must be run to produce outputs.

The application is a technical concept that aims to help teams find the application nesting the defective transformations in case of data issues. The application can be considered as the container of any data transformation; we will refer to this as a lineage entity. The application by itself is not sufficient to define what causes the data transformation. The application can evolve, which is why it is also important to log the code base and version of the application. Following the version of a script, different data sources can be used.

Code base

The code base corresponds to the location of the code that's running the application or other elements that can help you identify the repository of the code or the address of the program. For instance, the code base can be the URL of a Databricks notebook, the GitHub repository's location if you use a repository manager such as GitHub or GitLab, a Tableau workbook, and so on. The code base is related to an application, and an application can have several code bases. The same script can have several locations if it's run online (on a server) or offline (on the developer's machine) for instance.

The code base of our script can be found at `git@github.com:PacktPublishing/Data-Observability-for-Data-Engineering.git`.

The main goal of reporting the code base is to immediately spot where the observer has to look when they're debugging an application. The information is completed by the code version, which describes how the application evolves.

Code version

The code version represents an instance of the application. The version allows you to identify what exact version of the code was run when you were manipulating the data. Following the version of the application, several data sources may have been used. You can have your proper versioning system, such as v1.0.0, v1.1.2, and so on, or use a version control system such as git or even a timestamp. The code version is bonded to your code base as the same code base may have several versions.

The code version can also help you detect whether the version of the code is outdated or has not been updated for a long time, which may increase the occurrence of data issues if they're not aligned with the other applications.

The code version can also be linked to a user – that is, the one who created the code or the program version.

Our notebook adopts the commit hash as the code version.

Project

The project is the technical or business goal of a series of data transformations happening in one or several applications, also called the data pipeline. An example would be a data science project creating a recommendation engine. This project will consist of a sequence of different applications: a data extraction script written in Python, a data manipulation and transformation job written in Spark, a model prediction job using the Gluon Data API, and a Tableau workbook to show the results.

Running an application in a project allows you to understand the business or technical goal of the application, and how it interacts with other applications to fulfill that goal.

In this pipeline, you can imagine each application as if it is a brick that's used for building the project. However, the same brick can be used as the foundation for several projects. The Python data extraction script can produce a dataset that will be used by another reporting project. In this case, the owner of the second project would be interested in monitoring the data produced by the first application of the other project. Thus, the application belongs to both projects. This is illustrated in *Figure 4.2*:

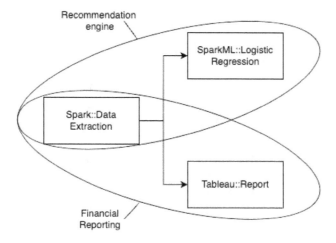

Figure 4.2 – Applications and projects

Here, we can see two projects: a recommendation engine and financial reporting. Both projects use the same Spark application as input. A change in this application will potentially impact both projects.

Here are some other examples of projects:

- A business process, such as a *client registration journey*. A business process is a sequence of actions (with data transformation and manual approvals) that have the same business goal.

- A technical process such as *ingesting and preparing the CRM*.

In our case, we will assume the notebook is part of a bigger project called *Company reporting*.

Environment

One of the best practices in IT is to work with several environments. CI/CD pipelines and their diverse environments are also part of the context of the data transformation. The same project and application can run in different environments, and the data may be different among them. It's an interesting parameter to observe, especially to compare run behaviors across environments. For instance, the data structure may be different in production compared to staging if some mismatch happens in the deployment that leads to some application or data issues.

Here are some examples of environments:

- Development
- Staging
- QA
- Pre-production
- Production

Our notebook is still at a preliminary stage. We will run it in a staging environment.

User

The user can be an identifier, an email address, an acronym, or simply a name that will refer to a machine, a cluster, an executor, or a person, depending on the company's standards on security.

The user will be useful not only for detecting the owner of a data source, an application, or a project, but also to know who wrote a piece of code, and who executed it.

For this example, we will assume the root user created the code and ran it on the machine.

Timestamp

To complete the execution context, it is standard to log the timestamp of the execution. The timestamped data can be collected when the application starts running and when it stops, or regularly when events such as data transformation take place.

The timestamp of the notebook's execution will be computed when the application starts running.

The application run

The application run is a concept that binds all the elements we've discussed so far. The run corresponds to the execution of an application.

A complete execution should at least answer the following questions so that an observer understands them:

- What application is being run?
- For which project(s) is the run being performed?
- In which environment is the application running?
- Who is running the application?
- What version of the code or program is running?
- Where can you find this version?

In an application run, you link the application and how it was run, providing you with an exhaustive contextual set of information.

As you may have noticed, we've encountered our first dynamic element of observability here. While other elements we've considered were quite easy to document, even if this required some discipline, dynamic events such as the application run are nearly impossible to trace without a good, automated logging strategy.

Also, the application run can be identified with an ID, a timestamp, or any information that would also allow us to connect the collected data observations with the world of application or infrastructure observation.

For each set of observations, we will show you how the Kensu log file creates and records them. Now that we have a good understanding of the context of the run, let's visualize it in the logs. The log file we'll create will be used to observe the data sources inside the applications and the pipeline. This file can be used to analyze issues when they happen so that we can prevent them and document the pipeline. This will be covered in the next few chapters. So, let's focus on the logs.

What's in the log?

To create a log file, you can run the first cell of the notebook:

```
from kensu.utils.kensu_provider import KensuProvider
K = KensuProvider().initKensu()
```

This cell calls the Kensu library and initializes a client. This client will store all the events we would like to log and will also create a file called `kensu_offline_events.txt` in the working directory.

At the time of initializing the client, while running the first cell, the context gets logged in the file. Let's look at the elements we have.

While running the cell, a file named `kensu_offline_events.txt` was created. The first cell of the notebook logs the context of the latter. Have a look at the file; we will describe some elements here.

One of the observability elements is `PROCESS`; this represents the application. Let's look at an important part of the JSON that was created:

```
{
        "action": "add_entity",
        "entity": "PROCESS",
        "jsonPayload": {
            "pk": {
                    "qualifiedName": "Compute_YTD"
            }
        }
}
```

The JSON that's created by the library, inside `jsonPayload`, contains `entity`, which is the observability element. In this case, the `PROCESS` entity contains a name (`Compute_YTD`) that is unique for the whole enterprise system, so it is considered the primary key (`pk`) of the entity. Similarly, the library file has created entities for `PROJECT`, `CODE_BASE`, `CODE_VERSION`, `USER`, and `PROCESS_RUN`. Here, `PROCESS_RUN` describes the execution details of the application. Here is an example of `entity`:

```
{"action": "add_entity", "entity": "PROCESS_RUN",
"generatedEntityGUID": "empty", "schemaVersion":
"0.1", "jsonPayload": {"pk": {"processRef": {"byGUID":
"k-258d3ac3637d631306c7716b2774430e659e67c8495aaee19f09850c6a224f2f"},
"qualifiedName": "Compute_YTD@2022-07-
05T19:35:41.934219"}, "launchedByUserRef": {"byGUID":
"k-6f3c2354458f395946b1861ec303725f40853600a6b000fc388a0d2d5d9c0caa"},
"executedCodeVersionRef": {"byGUID":
"k-344eb50279f7482a77c47df27e2141ab2d4b32662d45e86e03d332df77a180a4"},
"environment": "Staging", "projectsRefs": []}, "context": {"clientId":
"", "clientEventTimestamp": 1657042541935, "serverReceivedTimestamp":
1657042541935}}
```

As you can see, the PROCESS_RUN entity refers to the static elements such as the process, user, or code version. The timestamp of the run is saved in the name of the entity: Compute_YTD@2022-07-05T19:35:41. This will be the ID of the execution.

The different elements of observability that are set in the context are sometimes manually set in a configuration file or in the initializing function of the logging library, or automatically retrieved from the environment using variables.

The entities we've collected so far fit like this in our model:

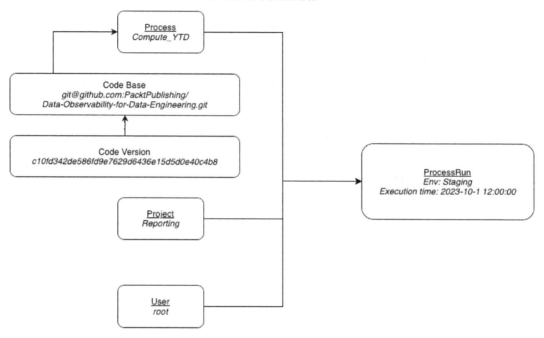

Figure 4.3 – The Compute_YTD application's run diagram

Now, let's look at the elements that are related to the data that's used inside the application.

Getting the metadata of the data sources

The fuel for the data application is the data itself. The data sources that are used in the application have to be correctly identified in the logs. If an issue occurs at a data source and you need to perform deeper analyses, you would expect information that will help you retrieve the data. In this section, we will see how a data source can be identified.

Data source

To identify the data that's used by an application, we need to define the metadata of the data source. The metadata represents the *data on the data*.

The metadata of the data source is all the elements that will allow you to recognize the data source. Let's explore them:

- **The file's location**: This gives you the address of the data source and helps you retrieve the data in case you need it. The file location can be the path on your local filesystem or the filesystem of the company. It can also be a connection string if the data is in a table located in a SQL server for instance. The path is something unique; you cannot have two different data sources with the same location. This is why the location is a unique identifier for the data.

- **Name**: A comprehensible name from a business perspective; for instance, the data source location may be `system://hd/data/BJT2023`, while the name could be `Customers_2023`.

- **Format**: Examples include a JSON file, a CSV file, a SQL table, a BigQuery table, and so on. This attribute can help you select the right strategy to query the data source so that you can analyze it.

- **Owner**: A name or an entity that's considered the owner of the data source and can be addressed in case of issues with the data.

- **Physical location**: Where the data is located. This adds documentation about the data and can allow for some **General Data Protection Regulation (GDPR)** monitoring.

In our script, we handle three different data sources, and all of them are CSV files. The data source is also the container of its schema(s). Let's see what it is in more detail.

Schema

The **schema** of the data source represents the structure of the content of the data. It is simply the set of columns that belong to a dataset. Each column, or field, can be defined with its properties:

- **The name of the field**: As it is represented in the data source column's name.

- **The type of the field**: This describes the kind of data contained in the column, for instance, string, double, or Boolean. The type can also be conceptual and gauge the conformity of the data (phone number, zip code, and so on).

- **Its nullable property**: A Boolean value that tells the observer whether the column can include missing values or not.

A single data source may have several schemas, and the schema itself can change over time. In the first case, an example is an Excel workbook. If you consider the workbook as a data source, each sheet can be a schema of the data source, with its own properties. In the second case, a change in the column type or the field name will automatically create a new schema associated with the data source.

While logging schemas, you may think about tracking the full schema or only the part that is used in the application. In big enterprise datasets, logging part of the schema is better. This will prevent you from logging hundreds of columns if only three are used for your specific use case. You will be more interested in knowing that the columns of interest are present in the dataset, rather than knowing the dataset is complete for all the use cases. This being said, many schemas can be created for the same dataset, none of them being complete, but each of them improving the observability of specific usage.

In our notebook, we also logged the schema at the time we read the data sources.

Once you have achieved this level of observability, you still need to link the applications and the data sources. Dependencies between the schemas are called lineages. This is the next element of observability we will cover.

What's in the log?

We advise you to log the data sources and schemas each time you read or write a data source. In kensu-py, logging is done when pandas reads or writes a data source.

Run the second cell to see the results:

```
import kensu.pandas as pd
df_jan = pd.read_csv("../data/jan/orders.csv",parse_dates=['date'])
df_feb = pd.read_csv("../data/feb/orders.csv",parse_dates=['date'])
```

Here, we are reading two data sources. In the kensu_offline_events.txt log file, you can find new observability elements: four new JSON lines representing the two data sources and their schema. Let's focus on the first one.

The first JSON was created for the DATA_SOURCE entity. It contains information such as its name, its format, the unique identifier, and its file location. To build the DATA_SOURCE entity, the library uses the first argument of the pandas function to retrieve the file's location. The format is inferred from the read function, and the name is based on the location:

```
{"action": "add_entity", "entity": "DATA_SOURCE",
 "generatedEntityGUID": "empty", "schemaVersion": "0.1", "jsonPayload":
{"name": "jan/orders.csv", "format": "csv", "categories":
 ["logical::orders.csv"], "pk": {"location": "file:/Users
data/jan/orders.csv", "physicalLocationRef": {"byGUID":
"k-d2f40e99e5dd4c9fc9c634b15a7fb03073191c0158e52a572769df8c05f59b7b"}}},
 "context": {"clientId": "", "clientEventTimestamp": 1657205496444,
 "serverReceivedTimestamp": 1657205496444}}
```

The schema that refers to this data source is represented by another entity. The schema contains a reference to the data source. Therefore, its unique identifier is made up of the data source reference and the list of fields defined by their name, type, and nullable properties. To compute the schema, the `kensu-py` agent used a DataFrame and performed a `dtype` method:

```
{"action": "add_entity", "entity": "SCHEMA", "generatedEntityGUID":
 "empty", "schemaVersion": "0.1", "jsonPayload": {"name":
 "schema:jan/orders.csv", "pk": {"dataSourceRef": {"byGUID":
"k-bb61b55fa3f40e0f41fb0ee084f667ce9937f8b0b586e99cb2962426ca69f280"},
 "fields": [{"name": "date", "fieldType": "datetime64[ns]", "nullable":
 true}, {"name": "order_id", "fieldType": "object", "nullable": true},
 {"name": "email", "fieldType": "object", "nullable": true}, {"name":
 "page_visited", "fieldType": "float64", "nullable": true}, {"name":
 "duration", "fieldType": "float64", "nullable": true}, {"name":
 "total_basket", "fieldType": "float64", "nullable": true}, {"name":
 "has_confirmed", "fieldType": "int64", "nullable": true}]}},
 "context": {"clientId": "", "clientEventTimestamp": 1657205496445,
 "serverReceivedTimestamp": 1657205496445}}
```

The data sources and schemas of the notebook are described in *Figure 4.5*. As we can see, three data sources are handled by the notebook – the two that are read and the one that is created:

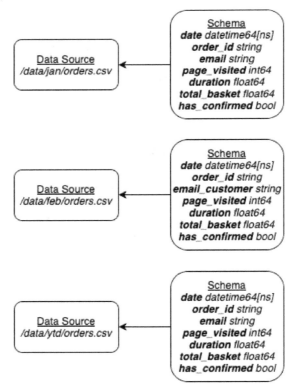

Figure 4.4 – Composited data sources and schemas diagram

As the schema refers to the data source, several schemas can point to the same data source. Schemas can be used to create or update another schema. This is what we will see in the next section.

Mastering lineage

Lineage or **process lineage** is the action of a data application on the data sources' schemas. Lineage is a link between inputs and outputs, often one or several input schemas and an output schema.

It expresses what happens with the data inside a specific application. By extension, the lineage of the data source is the set of all the transformations that ended in creating the data source and all the computations or manipulations that are based on the data source.

As we stated previously, lineage is a link between schemas. These schemas can come from the same data source. For instance, creating a new column inside a SQL table creates a new schema inside the table that is fed by data coming from another schema of the data source.

Lineage is a unique combination of data flows – a data flow being a one-to-one relationship between an input schema and output schema that occurs inside the application. Without the application, there cannot be any lineage. The application is the engine that manipulates the data and creates the links among sources. Therefore, lineage is a combination of input(s), output(s), and an application. Here are some examples:

- A Python script that reads a CSV file and inserts its data into a relational database
- An ETL process that extracts data from multiple APIs and aggregates it into a unified data format for further analysis
- An AWS Lambda function that triggers when an S3 file is uploaded and processes the file to generate summary statistics

Let's dig into the different types of lineages.

Types of lineage and dependencies

Lineage can be computed at several levels:

- **Data source-level lineage**: At the data source level, we focus on the relationship between the tables or data sources. For instance, you know that CRM is used to create a customer view, but you don't have more details about how the original data source is used. Nevertheless, this is a prerequisite for field-level lineage.
- **Field-level lineage**: Field-level lineage is about knowing how the output data source was created. It adds another level of detail to the lineage. You know that the customer view is coming from CRM, and you learn from the field lineage that the `Full_Name` column in the customer view comes from the concatenation of two columns in the CRM – the first and last names. Depending on the complexity of the technology, it may be complicated to handle field-level lineage.

There are two kinds of dependencies between fields inside the lineage concept – column and control lineage:

- **Column lineage** is when the relationship between two data sources is based on data items being exchanged between the sources. The data item of the input data source is transferred to the output end of the lineage, whether the data is modified or not. It's the equivalent of a SELECT statement in SQL. The result contains data from the original table.

- **Control lineage** is a relationship between tables or fields where no data is exchanged, but the result of the operation depends on the data contained in the inputs. For instance, it can be a WHERE or ORDER BY statement in a SQL query. The data is not read nor transferred to the output; however, a change in the inputs can lead to unexpected changes in the outputs.

Lineage is considered a static element; to make it dynamic, it has to be run. So, let's explore the concept of lineage run.

Lineage run

A **lineage run** is to lineage what an application run is to an application. It links the lineage to the right application run and tells the observer when the transformation was made.

Inside the same application run, you can have the execution of many lineages. Some of them can be played more than once. Think about a streaming application. The application is launched once. This is the start of the application run. Inside the application, following what arrives in an S3 bucket, the application will read the file and copy it to another storage place, creating a lineage. If the same file is reuploaded, the same lineage will be rerun. You will have two lineage runs for the same application run.

What's in the log?

The logs of the lineage are produced when a data source is written. Once done, the Kensu client will analyze all the events that led to the creation of the data source.

First, let's describe the logic that was used to create the lineage. In our script, the February order data source schema was modified by changing the name of the email_customer column. Then, both the January and February sets were merged and saved in a new file:

```
df_feb = df_feb.rename({'email_customer':'email'},axis=1)
data_YTD = pd.concat([df_jan,df_feb])
data_YTD.to_csv("../data/feb/ordersYTD.csv",index=False)
```

While doing these programmatic steps, the kensu-py agent did the following:

1. It logged that the name of the df_feb column, email_customer, was renamed email.

2. It also logged that the data_YTD DataFrame was created out of df_jan and df_feb with a concat pandas function.

3. When writing the DataFrame to the filesystem through the to_csv method, the agent computed the field-level lineage by associating the list of events that took place in the execution.

The result is a PROCESS_LINEAGE entity in the logs. Here are the core pieces of the entity, namely its list of data flows:

```
"dataFlow": [{"fromSchemaRef": {"byGUID":
"k-e42c954f0243efb90efa144606da577d3c7e18d8bb0941cd2d5e88f7b5c0c6dc"},
"toSchemaRef": {"byGUID":
"k-85c2fd35707da2f27b7ab69c77b422b5d34aeee5a75b061b1cbf0bbc40077640"},
"columnDataDependencies": {"date": ["date"], "duration": ["duration"],
"email": ["email"], "has_confirmed": ["has_confirmed"], "order_
id": ["order_id"], "page_visited": ["page_visited"], "total_
basket": ["total_basket"]}}, {"fromSchemaRef": {"byGUID":
"k-7fd5a9f6ebc509ae2e626edf716cf67638dd0645654cbf356ef6a3f97a49efb4"},
"toSchemaRef": {"byGUID":
"k-85c2fd35707da2f27b7ab69c77b422b5d34aeee5a75b061b1cbf0bbc40077640"},
"columnDataDependencies": {"date": ["date"], "duration": ["duration"],
"email": ["email_customer"], "has_confirmed": ["has_confirmed"],
"order_id": ["order_id"], "page_visited": ["page_visited"], "total_
basket": ["total_basket"]}}]}}
```

The data flow represents the list of data dependencies among the input and output schemas. The items in the list are one-to-one relationships – they bind each of the inputs to the output.

Inside each data flow, you can observe the columnDataDependencies key. This is the field-level lineage. For each field of the output data source's schema, it lists the input fields that were used to create, update, or fill the output column. The second data flow contains "email": ["email_customer"], which signals that the email_customer field of the February dataset was used to fill the email column of the output dataset.

The logs also contain a LINEAGE_RUN entity. This lineage run binds the previously created PROCESS_LINEAGE to PROCESS_RUN and adds the timestamp of the lineage's execution. Thanks to this, you know the context of the execution of the data transformation.

Here's a diagram of this lineage:

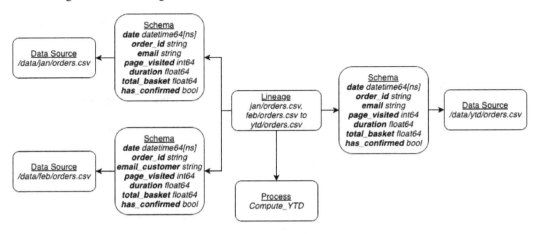

Figure 4.5 – Compute_YTD lineage diagram

For ease of understanding, we have separated the inputs on the left from the output on the right. The two data sources, from the January and February folders, are joined through the Compute_YTD process and create the ytd/orders.csv data source.

The lineage run is represented in *Figure 4.6*. As you can see, the timestamp of the application run can be different from the timestamp of the lineage run. The first timestamp defines the moment the application started, while the second one is defined by the moment the data transformation is applied and the final data source is written:

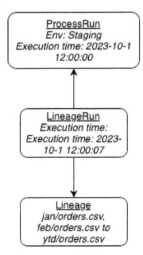

Figure 4.6 - Compute_YTD lineage run diagram

A bunch of other entities are also computed at the time of writing. We'll discover them in the next section.

Computing observability metrics

The following data observability elements are known as **data quality metrics**. In this category, we will group everything we consider to be observability metrics. These observations are statistics related to the data you manipulate:

- **Distribution observations**: Minimum, maximum, mean, standard deviation, skewness and kurtosis, quantiles, and so on

- **Categorical stats**: Number of categories, percentage of each category, and so on

- **Completeness observations**: Number of rows and number of missing values

- **Freshness information**: Timestamp of the data itself

- **KPIs**: Key performance indicators and other custom metrics worth checking, for technical or business purposes

The metrics you compute depend on the circumstances and need to be linked to the context where they were computed. Those metrics can change following the usage of the data, the filters you applied, and the application run. *Figure 4.7* shows an example of multiple contexts for the same data sources:

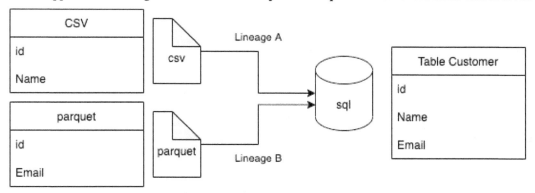

Figure 4.7 – Example of a data source created by two applications

In the preceding figure, we can see two different operations being performed to add data to the SQL database. The first one, **Lineage A**, adds the **Name** column to the table, while the second one adds **Email** information. Likely, you will compute the number of rows in the CSV and Parquet files, and the number of affected rows in the customer table, to avoid a huge load on the database, especially in cases when the central customer table contains terabytes of data and hundreds of columns. Thus, the number of affected rows will depend on the data you have in the inputs.

The statistics are dynamic elements and need to be linked to the lineage run and the end of the lineage – that is, the schema. Each execution of the lineage can bring new statistics, and those are nested in the schema of the right data source.

Let's see what this looks like in practice.

What's in the log?

When writing the data source to the filesystem, besides the lineage and lineage run computation, the client also produces the statistics of all the data sources that are used in the lineage:

```
{"action": "add_entity", "entity": "DATA_STATS",
"generatedEntityGUID": "empty", "schemaVersion":
"0.1", "jsonPayload": {"pk": {"schemaRef": {"byGUID":
"k-e42c954f0243efb90efa144606da577d3c7e18d8bb0941cd2d5e88f7b5c0c6dc"},
"lineageRunRef": {"byGUID":
"k-3e8bb4c9bdfb1bd2b687c1103843b85db47b491b61db5a054a5e52c571369447"}},
"stats": {"page_visited.count": 331.0, "page_visited.mean":
8.024169184290031, "page_visited.std": 3.961866198819761, "page_
visited.min": 1.0, "page_visited.25%": 5.0, "page_visited.50%":
8.0, "page_visited.75%": 11.0, "page_visited.max": 19.0, "duration.
count": 359.0, "duration.mean": 537.5388300835655, "duration.std":
289.5523807964363, "duration.min": 40.07, "duration.25%": 290.195,
"duration.50%": 541.65, "duration.75%": 787.55, "duration.max": 996.73,
"total_basket.count": 359.0, "total_basket.mean": 423.2941225626741,
"total_basket.std": 223.96462692755878, "total_basket.min": 15.11,
"total_basket.25%": 228.79000000000002, "total_basket.50%": 423.74,
"total_basket.75%": 615.81, "total_basket.max": 782.77, "has_confirmed.
count": 359.0, "has_confirmed.mean": 0.8662952646239555, "has_confirmed.
std": 0.3408098009102848, "has_confirmed.min": 0.0, "has_confirmed.25%":
1.0, "has_confirmed.50%": 1.0, "has_confirmed.75%": 1.0, "has_confirmed.
max": 1.0, "date.count": 359, "order_id.count": 359, "email.count": 359,
"date.first": 1609459200000.0, "date.last": 1612051200000.0, "nrows":
359, "date.nullrows": 0, "order_id.nullrows": 0, "email.nullrows": 0,
"page_visited.nullrows": 28.0, "duration.nullrows": 0.0, "total_basket.
nullrows": 0.0, "has_confirmed.nullrows": 0.0}}, "context": {"clientId":
"", "clientEventTimestamp": 1657208447483, "serverReceivedTimestamp":
1657208447483}}
```

Here, you can see the statistics that are linked to columns of the schema of the first data source. Numerical data statistics, including the mean and the distribution metrics, were computed using the pandas DataFrame .describe() method. For those indicators, quality targets can be set. We will cover this in *Chapter 5*.

The date column was converted into EPOCH to compute the date of the oldest and the most recent data rows. This was done so that the data model can consider the value as a numerical value and not as a timestamp format. This can be used to compute the freshness of the dataset.

Several metrics are computed on the output data source:

nrows	page_visited				date		
	min	max	mean	std	First	Last	nullrows
747	1	20	8.21	3.94	1609459200000	1614470400000	0

Table 4.1 – Metrics based on the output data

Figure 4.8 shows the next part of the diagram binding the DataStats entity with LineageRun and Schemas:

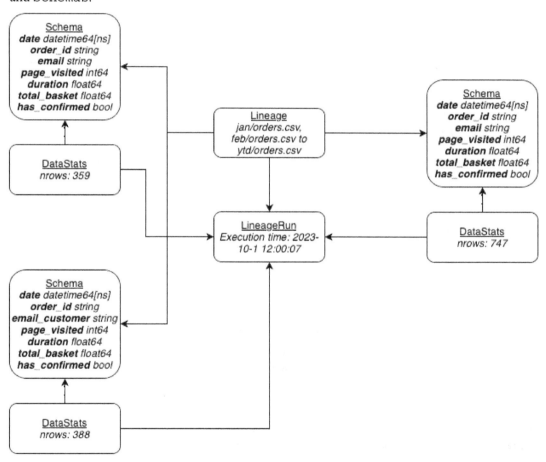

Figure 4.8 – Compute_YTD DataStats diagram

In this section, we looked at the many elements of data observability for engineers. The following section will explain how this data observability model can be extended to cover data scientists' and analysts' applications.

Data observability for AI models

Here, we would like to mention some specific elements of data observability that focus on AI and ML methods. You can see the model as a particular case of a data source. The model is the output of a lineage. For this, you can follow the second notebook for this chapter, `Orders_predict`.

Let's look at the different components of ML observability.

Model method

The model method is the name of the method that we will use to apply the transformation to create the model. It will be, for instance, the name of the scikit-learn classes you use or a more generic method name:

- **Example of a library method**: `Scikit::LinearRegression()`
- **Example of a generic method**: Random forest

The method is the ingredient you use to create the model data source. Inside the same application, you can try several methods and compare them. At this point, you must link the method to the right lineage. To do so, you must use the model training entity.

Model training

The model training will link the model method to the correct lineage operation. It signals that the lineage is the application of a model method. The same model can be trained several times on the same data source, but rows can be different (that is, a different size of training set).

The model training process plays an essential role in linking the model method to the correct lineage operation, which is crucial for maintaining traceability and understanding the model's development history. The lineage operation serves as a record of the model method's application, capturing the specific configuration and parameters that are used during the training process. This enables data scientists to trace back the model's performance to the training settings, providing valuable insights into the factors affecting the model's behavior.

When the same model is trained several times on the same data source, it's common to have different training set sizes or row selections. This variability can be due to several reasons, such as different data sampling methods, data partitioning techniques, or updates to the data source over time.

By recording the details of each model training run, the lineage operation allows you to compare different model configurations and their corresponding performance metrics. This information allows data scientists to experiment with various training set sizes, data preprocessing techniques, and model hyperparameters to optimize the model's performance. Furthermore, they can determine which factors significantly impact the model's performance and focus their efforts on addressing those issues.

Model metrics

The model metrics are to the model data sources what the statistics are to the genuine data sources. The model metrics are linked to each run of the model training lineage. A new run of the lineage could generate new metrics if the content of the data source is different.

What's in the log?

The following code example shows the log of the model:

```
import kensu.pickle as pickle
from kensu.sklearn.model_selection import train_test_split
import kensu.pandas as pd

from kensu.utils.kensu_provider import KensuProvider
KensuProvider().initKensu(allow_reinit=True)

data = pd.read_csv("orders.csv")
df=data[['total_qty',  'total_basket']]

X = df.drop('total_basket',axis=1)
y = df['total_basket']

X_train, X_test, y_train, y_test = train_test_split(X, y, test_
size=0.3, random_state=0)

from kensu.sklearn.linear_model import LinearRegression
model=LinearRegression().fit(X_train,y_train)

with open('model_copy.pickle', 'wb') as f:
    pickle.dump(model,f)
```

In this example, a model is being trained on the `orders.csv` dataset and then saved in the form of a pickle file.

The log file that was produced in this script contains several pieces of information:

- The metadata of the input `orders.csv` data source, including its schema

- The output, which is `model_copy.pickle`

- A lineage that combines references from `orders.csv` to `model_copy.pickle`

- Additional information about the model, which is described in this JSON:

```
{"action": "add_entity", "entity": "MODEL", "generatedEntityGUID":
"empty", "schemaVersion": "0.1", "jsonPayload": {"pk": {"name": "SkLearn.
LinearRegression"}}, "context": {"clientId": "", "clientEventTimestamp":
1682028035162, "serverReceivedTimestamp": 1682028035162}}
```

First, we log the model method, which is a `SkLearn.LinearRegression` model:

```
{"action": "add_entity", "entity": "MODEL_TRAINING",
"generatedEntityGUID": "empty", "schemaVersion":
"0.1", "jsonPayload": {"pk": {"modelRef": {"byGUID":
"k-421022d34f0ce88ad61bbb1db30f674a34702bbd7f0f8e513f6d4e1502cdb471"},
"processLineageRef": {"byGUID":
"k-db6afd8b8e1a98a8e3d13df5ac608b52c009b2583c0bb7ee13576f6230001199"}}},
"context": {"clientId": "", "clientEventTimestamp": 1682028035162,
"serverReceivedTimestamp": 1682028035162}}
```

Second, we signal that the lineage that's used in this application belongs to the model training category by referencing `PROCESS_LINEAGE` and `MODEL`:

```
{"action": "add_entity", "entity": "MODEL_METRICS",
"generatedEntityGUID": "empty", "schemaVersion": "0.1",
"jsonPayload": {"pk": {"modelTrainingRef": {"byGUID":
"k-1d65aae773dbf39084224e5e3181618de4377adef5c6477efa90217664325e28"},
"lineageRunRef": {"byGUID":
"k-d20191f533d5ff9c32c153512c2bd40ce71ce47ff8d2e7e0a8e7497065ebe729"},
"storedInSchemaRef": {"byGUID":
"k-2385ce655bdd0639d9fbb8ffa6cc0782be28ea895c5cfe44a5a421ac437511f1"}},
"metrics": {"train.explained_variance": -63.57729809163885, "train.
neg_mean_absolute_error": 113.88108183295255, "train.neg_mean_squared_
error": 16816.822186619163, "train.neg_mean_squared_log_error":
0.6259570196657532, "train.neg_median_absolute_error": 119.09181687428375,
"train.r2": -63.57729809163885}, "hyperParamsAsJson": "{\"copy_X\":
true, \"fit_intercept\": true, \"n_jobs\": null, \"normalize\":
\"deprecated\", \"positive\": false}"}, "context": {"clientId": "",
"clientEventTimestamp": 1682028035162, "serverReceivedTimestamp":
1682028035162}}
```

Finally, we log some metrics about the training performance. This observation includes the following:

- Metrics such as explained variance, negative mean absolute error, negative mean squared error, negative mean squared log error, and the R2 score.

- The hyperparameters of the model when it was trained and when it was retrieved with the `sklearn get_params()` method

With another model method, we could have gathered other metrics, such as the number of true positives, false positives, true negatives, and false negatives, and derived metrics such as the recall or F1 score.

The model metrics and data science considerations were not presented in the generic model shown in *Figure 4.1*.

Figure 4.9 shows the new entities and their dependencies on the rest of the model:

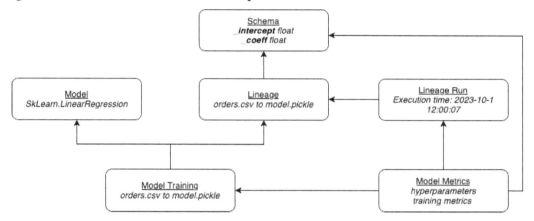

Figure 4.9 – Orders_predict metrics diagram

These metrics are important for us to understand how the model behaved when it was created. When the model runs in production, the training metrics can be confronted with reality by what is called a feedback loop.

The feedback loop in data observability

In data observability, a feedback loop is an essential component for the continuous improvement of machine learning models. By leveraging the information gathered during both the training and production phases, a feedback loop enables developers and data scientists to monitor, evaluate, and optimize their models for better performance.

During the training phase, model metrics entities play a crucial role in establishing an initial understanding of the model's expected performance. These metrics, such as the accuracy, precision, recall, and F1 score, help the observer gauge how well the model is likely to perform on unseen data. By analyzing these metrics, data scientists can identify areas of improvement and fine-tune the model accordingly.

Once the model has been deployed in production, data stats entities come into play. These entities allow you to compute various metrics by comparing the model's predictions against the actual data when it becomes available. This real-world data is invaluable for assessing the performance of the model in its intended environment.

Incorporating the insights from data stats entities into the feedback loop allows for a better understanding of potential discrepancies between the model's predictions and reality – that is, the common `y_pred` and `y_true`. This information can be used to diagnose issues such as data drift and model degradation. Additionally, it can highlight the need for retraining the model or adjusting its parameters to maintain optimal performance.

By continually monitoring and analyzing the information from both model metrics and data stats entities, data scientists can iterate on their models and refine them based on real-world feedback. This continuous improvement process ensures that machine learning models remain accurate and reliable, adapting to changes in the data landscape and meeting the evolving needs of the end users.

Summary

In this chapter, we covered the important elements that we need to collect to implement observability at the data level from within the application. This observability was exposed in a data model, where we distinguished several categories of observations.

First is the elements related to the context – that is, what application is running, what version it is using, who created it and who runs it, and where and when it was run. These elements are important to create a structure around the data transformations. Second is the data itself. We saw that the metadata can be defined by some attribute of the data source and its schema. Third are the data transformations and operations, which we have described as lineages. These lineages are also the link between the data sources, their schemas, and their applications. Finally, once we have associated the lineage with the right execution, some observation metrics can be computed.

We also looked at some specific elements related to the AI/ML world, and we looked at their models and metrics. Using the `kensu-py` library, we applied monkey patching to a Python `pandas` script to generate a log file containing the previously mentioned data observability elements. The produced logs are the basis of the documentation of the pipeline, which will result in root cause analysis and prevention.

All these elements can also be the base of indicators. The observation elements described in this chapter will be the origin of statistical rules, which can enhance the quality and health of the data source. We will cover this in the next chapter.

5

Defining Rules on Indicators

In the previous chapters, we saw how you could collect events synchronously in your data applications. We also discussed what contextual information you need in order to draw the big picture of what's happening inside the applications.

Now that you have a lot of contextual information, it is high time to turn it into actionable insights. The metrics you collect during the pipeline execution need to reassure all the stakeholders about the proper execution of the data applications. All the observers of the pipeline need to be informed about how the data pipeline is behaving.

To maintain the trust of data producers and data consumers, we will introduce the concept of *expectations*, which will define what the engineer needs to achieve in order to keep the pipeline in good shape. These expectations, composed of metrics and rules, will act as sensors to know whether the applications are working as expected or not.

These rules are a key component of data observability as they will allow us to focus on the issues by doing the following:

- Reducing the time needed to detect anomalies
- Providing a clear context of the data issue to stakeholders
- Preventing issues and activating continuous validation

Throughout this chapter, we will learn how the **service level objectives** (**SLOs**) of the data can be put in place, first by describing methods to create them at the project and data source levels, and then by explaining how these can be implemented inside the enterprise pipelines. We will see how the indicators collected during the application runs can be used to follow objectives.

At the end of the chapter, we will also cover how this validation can be switched on in a CI/CD pipeline, allowing continuous validation of your pipeline, a best practice for a well-observed pipeline.

The topics that will be covered are the following:

- Determining SLOs
- Turning SLOs into rules
- Project – continuous validation of the data

Technical requirements

For this chapter, you would use the same Python environment created for *Chapter 4: Data Observability Elements.*

Determining SLOs

We have already described the SLA-SLO-SLI paradigm in *Chapter 2, Fundamentals of Data Observability.* The agreements are individual (implicit or explicit) contracts between a producer and a consumer that set the requirements that the produced data has to meet in order to be considered healthy. The objectives are targets that the producer must meet in order to fulfill their agreements. Finally, indicators are means of gauging whether the objectives are respected or not.

But first, let's see where SLOs can be added.

Project versus data source SLOs

The objectives can be set at different levels, depending on the needs of the stakeholders. Indeed, some objectives can be set at the project or pipeline level, and others at the data source level. You can see this at the micro and macroscopic level. Depending on the scope of responsibility you have, you may need to set the objective at one level or the other.

At the microscopic level, the objectives set on a project will often cover a single agreement, while at the macroscopic level, objectives will try to fulfill as many agreements as possible, and this is because the data source may be used in several projects.

Combined with observability, it becomes easier to understand what impact a data issue will have. In general, if issues impact all the projects, then we will create rules at the data source level. Specific needs for some projects will be included in the project's objectives.

When it comes to determining the objectives, it can be exhausting and difficult to find the right targets that will satisfy all the agreements. In this section, we will reveal the methods you can use to define your objectives at the data source and project level. We will subsequently explore four methods:

- Knowledge-based
- Starting from **service level agreements (SLAs)**

- Starting from **service level indicators (SLIs)**
- Inferred from past issues

Knowledge-based method

The first method is an easy one – you know what needs to be included in your project or database to make it compliant. This method is very powerful when it comes to technical constraints on your own data pipeline. By technical constraints, we mean those involved by the components of your data pipeline technologies or tools. For instance, if a SQL data source you feed does not accept any NULL value for a certain column, you can set an objective on the completeness of the input data. With this, you ensure the quality of your (part of the) pipeline. These rules can be included while designing the pipeline.

Another way of creating these rules is to trust your gut feeling! Indeed, a lot of rules can be included based on knowledge you can infer from the data. If you see a column called `Adult_Age`, you can assume that you would not see any ages lower than ~20 (depending on the country's legislation), and the values cannot be extreme (let's say, no more than 120). Even if those rules are bad assumptions, the events created by the violation of those rules would help the team distinguish the best rules that should keep out the noise.

This method comes of course with some limitations, the first one being that some created objectives are based on assumptions that may turn out to be inaccurate. As stated before, this can result in a lot of false positive issues that will require the team's attention for little return. Also, we have to keep in mind that these rules were not created based on business or consumer inputs, which can lead to different interpretations by the producers.

To avoid relying solely on your knowledge, you can also ask for some help from the consumer side. This is the purpose of the next method.

Starting from SLAs

This top-down approach consists of starting from the needs and wants of the business or the consumers to find out what is worth being observed.

As stated in the previous chapters, agreements are a good starting point to define the objectives. The objectives at the data source level must cover as many individual consumer-producer agreements as possible.

This source of objectives involves recurrent reassessment of the defined KPI, as the number of consumers but also their needs can vary over time. Also, the type of consumer can impact the definition of the objective. The consumers can be a good source if their level of knowledge of the pipeline or data product allows you to correctly translate their needs into metrics. However, it can also be a nightmare for the consumer as well as the producer to define interesting KPIs based on high-level agreements.

The SLAs of the projects are intimately linked to the context dimension of data observability. For each context, you can define an agreement and several agreements will be part of an objective, so that if an objective is not met, all the corresponding parties of the underlying agreement can be contacted.

In addition to this top-down approach, we can also have its bottom-up counterpart. Let's explore that in the following section.

Starting from SLIs

In the SLA-SLO-SLI paradigm, we have seen how we can use the needs of consumers in order to define the SLOs. These objectives can even be defined *ex-ante*, which means before the pipeline has been developed by the engineer. The indicators themselves can also help in defining those objectives once the pipeline has been developed, so in an *ex-post* approach.

Indeed, data observability methods require the collection of many contextual metrics (for example, the number of nulls, minimum, maximum, average, standard deviation, number of categories, frequency, and so on) about your pipelines and data sources. All these metrics can help you identify the normal behavior of your applications and thus, the important metrics to follow.

The main issue here is to reduce the noise introduced by the collection of diverse and numerous metrics. This said, some anomaly detection methods can help you in this process. We will cover those methods in *Chapter 6, Root Cause Analysis*.

Inferred from past issues

Past troubleshooting reports and their follow-up actions are also great sources to define objectives. They help to focus on what metrics may have been overlooked when defining the objectives.

The past issues method can help to refine the current objectives by changing their definitions or adapting them to new needs that have been overlooked – for instance, a new or modified agreement.

We will cover how this troubleshooting can be initiated in *Chapter 6, Root Cause Analysis*.

Use case

In this section, we will introduce a small use case to put into practice the notions we have seen on how we can set objectives on a specific data source.

Let's imagine the following situation: you are a data engineer in a telecommunications company. You are in charge of providing the other team with a *Customer 360 view* table in the central PostgreSQL database. This table is then the fuel of two data processes performed by another team:

- **Churn**: A model that identifies which customers are about to break the contract in order to take preventive actions

- **Loyalty**: A process where the consumers are ranked in order of their payments following their seniority and are granted some privileges

To create this view, you develop a small Python application reading various files coming from the central CRM:

- **Customer table**: Information about the customer (personal and sensitive info)

- **Contract table**: Information about their contract (telecom plan, seniority)

Let's see which objectives can be set at the *Customer 360 view* data source level.

Knowledge-based method

First, based on the application you are responsible for, you can infer your proper objectives for this part of the pipeline. As you are creating a table based on all the customer data you have in the CRM, you must be sure the process does not omit any customer or contract from the CRM. This is the first objective, on *completeness*.

SLA method

To find what KPIs are interesting to follow for the concerned data source, you need to start from the two usages, modelized here in projects, where it will be a source. A frank discussion with the concerned teams is needed. You can see the agreements you could define in *Figure 5.1*:

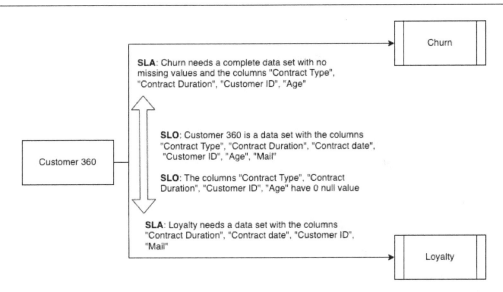

Figure 5.1 – Example of an SLO based on the SLA

The churn team has listed its needs. As it is using a data science model, it needs a complete dataset without any missing values. The team requires the maximum number of null rows in the columns of the dataset used in the model to be 0.

The loyalty team has other needs. Like the churn team, it will require a specific subset of columns to be present in the dataset. However, because the loyalty team doesn't have any strong technical constraints, it is not worried about any missing values.

You now have defined the needs of the consumers and can prepare several objectives. First are two global SLOs at the data source level, including the list of fields the data source must contain and the KPI that needs to be respected. Besides this, you can ensure the viability of your individual agreements by adding project-level objectives. These objectives will be strictly based on a single agreement and will help you to contact the right team in case the objective is violated.

SLI method

Once the data teams have prepared the data processes, the collected metrics can be used to define new objectives. For instance, the performance of the churn model is influenced by the standard deviation of the features. If you detect a high variability in these standard deviation metrics, it would be interesting to track it by adding an objective to ensure the viability of the model.

Past issues method

After some time running in production, the loyalty team comes back to you because of the *poor quality* of the dataset. It is experiencing an issue with too many customers being granted *diamond* privilege while this is only offered to users with more than 10 years of seniority. This results in a higher budget being allocated to the loyalty program. After some investigation, the loyalty team discovers that the `Contract Duration` column definition it used to infer the seniority of the customer has changed. The duration is now expressed in months instead of years, leading to a shift in the data distribution and a higher number of high-seniority customers because the data pipeline was not adapted.

As a follow-up, you can start to set an objective on the variability of this duration column in order to foresee any future issues linked to this.

Now that we have seen how objectives could be set in a pipeline, we will describe how and where they can be applied in the pipeline.

Turning SLOs into rules

In this section, we will see how objectives can be turned into actionable rules by creating contextual checkpoints from the pipeline or externally. At the start of any rule is the *expectation*, which can be defined as "*What does the consumer expect from the dataset?*"

An expectation formalizes the objective into a rule and the corresponding metric to be tracked. The expectation is then a good way to document the objectives and the metrics needed to respect them. The two components of the expectation have their importance: the rule tells the observer how the data should behave, and the metric is used to detect whether the behavior is deviant or not.

Let's look at the different types of rules that we can set.

Different types of rules

The backbone of a rule is the indicator. Based on this, a rule can be set and will start checking how the metric is behaving. These rules are often guided by the principles of data quality discussed in *Chapter 1*, *Fundamentals of Data Quality Monitoring*.

To keep it as simple as it should be, and to help the engineer in the rule definition at the code level, we advise creating as many rules as the number of cases you want to cover. The result of the check must be clear and unambiguous.

Rules can be set based on one single observation or on a time series of observations. Also, a rule can be directed by a single indicator or multiple indicators.

Here are some examples of rules that can be applied to numerical indicators:

- **Range**: The indicator cannot be higher or lower than some reference values.

  - Example: A *Minimum of Adult_Age* metric cannot be lower than 18

- **Variability** (over time): The indicator cannot vary by a certain threshold compared to a reference. This reference can be the last available observation or a subset of past observations.

  - Example: *Mean of Total Revenue* cannot vary by more than 10%

- **Comparison**: The indicator cannot be higher or lower than another indicator.

  - Example: *Total Number of orders* has to be higher or equal to *Total number of customers*

- **Freshness**: The data cannot be outdated when processing it.

  - Example: *Customer timestamp* cannot be older than last month

Other rules can also be added to non-numerical indicators:

- **Schema**: The data source must contain expected fields
- **Format**: The data source must respect a required format

This list, although not completely exhaustive, covers most of the requirements necessary for numerical indicators.

Implementation of the rules

Once the rules have been defined, they can be implemented at different levels:

- In an external system; the system oversees the collected observations and matches the rule to validate it
- Inside the code itself, as a condition or a circuit-breaker

Let us now look in detail at how we can implement and best use these two types of approaches.

External validation

This method consists of checking the validity of the rule like an observer would do. Based on the collected observation metrics, the rule engine will check the validity of each of the defined rules and create anomalies based on the results of those checks.

This external tool also allows documentation about the expectations of the data source by creating a register of all related rules.

In-code validation

The rules can also be added to your code in different ways. First, they can be added as preconditions before processing the data so you can avoid creating or propagating any data issues. Second, the result can be checked, even before writing the outcome to the filesystem, which means that preventive action can be taken based on the validation.

Conditional checks

The first form of in-code validation we will explore is a check of the data received and produced along the pipeline. At each step of the pipeline, the developer can validate the inputs and outputs to avoid garbage in, garbage out.

Following this validation, some decisions can be taken. Following each of these choices, troubleshooting actions can also be taken. The developer can choose to do nothing, or at least log the issue for the record. The developer can also introduce a logic to avoid the code continuing to run, called a **circuit-breaker**. Let's discuss this more in detail.

Circuit-breaker

A circuit-breaker in data stops the creation of the problematic data pipeline source by triggering an error so that the execution of the application is simply stopped.

Implementing these rules at one or another level can be driven by several factors:

- **The cost of doing nothing**: Writing an erroneous table may incur costs. When it comes to big data, the cost of storage or the resources needed to write the data source may be high, leading to the decision to include a circuit-breaker to avoid these costly operations.

- **The criticality of the error**: Some projects may be run even if the data is erroneous because the error is not impacting all the outcomes of the project. At this stage, a simple notification to the concerned team is enough, with no need to troubleshoot.

- **The number of dependent projects**: Does the error concern a single or a limited number of projects compared to the whole set of projects using the data source? To answer this, the observability context is a great ally and will allow you to target the communication needed to the consumers.

Let's see how this can be put into practice inside some simple code.

Example

The following code sample performs a simple operation. You can find the notebook in the GitHub repository (as a reminder, it can be found at `https://github.com/PacktPublishing/Data-Observability-for-Data-Engineering`).

The notebook is located in the `Chapter5` folder and is called `Data_Observability_Rules.ipynb`.

This piece of code aims at creating a register of the active users of a platform:

```
Name = pd.read_csv('Name_Surname.csv')
Age = pd.read_csv('Age.csv')
Personal_data = Name.merge(Age, on='id')
Personal_data.to_csv('Personal_data.csv')
```

It reads two CSV files and merges them based on the `id` column. The first file contains the name and surname of the user, and the second file contains the age of the individual. Both files contain the `id` column, which is the key to performing the join operation.

Some rules have to be implemented to support the viability of this script, and thus, the creation of the `Personal_data` file.

First, the two input files, `Name_Surname.csv` and `Age.csv`, must contain the `id` column or the process will fail. We then decide to create a checkpoint that will check the validity of the input. This can be done by adding the following code:

```
if 'id' not in Name.columns or 'id' not in Age. columns:
    logging.warning('Missing key in the input file(s)')
```

These lines of code will check whether the `id` column is well-defined in both input files to be sure the merge operation can be performed. This line is not a circuit-breaker. However, the code will stop if the logline is displayed because the merge operation cannot be performed correctly. However, thanks to the warning, the troubleshoot will be faster.

On the output side, we will add a circuit-breaker to avoid creating a wrong data source that may be assumed as correct by the consumer. The check we want to perform is to ensure that the created source has the same number of rows as the `Name_Surname` data source, considered as the reference. To do so, we will use the `check_nrows_consistency()` function of the `kensu-py` library. This function will check the number of rows in the inputs and throw an exception if the number of rows in the result data source is lower than expected. This is what the circuit-breaker looks like:

```
kensu.utils.exceptions.NrowsConsistencyError: Personal_data.csv has
less rows than expected: 10 out of a maximum of 19 - 53.0%
```

In this example, while there were 19 users in the `Name_Surname.csv` file, only 10 were matched with a corresponding `id` in the `Age.csv` file, leading to this `NrowsConsistencyError`.

Example with our use case

Back to the telecommunications example provided in the *Determining SLOs* section. Once the objectives have been defined, it is time to convert them into rules that will monitor the pipeline and report to the observers.

The objective here is to validate the completeness of the result table. The Customer 360 table is the result of the `join` operation between other tables. To ensure the data table is complete, the first objective you want to include in your observability framework is to not omit any customer or contract from the CRM. To translate this into a rule, you first need the following indicators:

- The number of rows of the customer table of the CRM
- The number of rows of the resulting Customer 360 table

Once you have these metrics, the rule can be formulated by the following:

Number of rows of Customer 360 table must be equal to number of rows of customer table

As this part of the pipeline is under your scope of responsibilities, you have decided to include a circuit-breaker. This will prevent the creation of any data source containing fewer or more rows than the customer table.

Similarly, the other objectives can be defined as follows:

Objective	Indicator	Rules
Customer 360 is a dataset with the columns "Contract Type," "Contract Duration," "Contract date," "Customer ID," "Age," and "Mail."	The schema of Customer 360.	"Contract Type," "Contract Duration," "Contract date," "Customer ID," "Age," and "Mail" are in the schema of Customer 360.
The columns "Contract Type," "Contract Duration," "Customer ID," and "Age" have 0 null values.	The missing value counts for the mentioned columns.	Four rules for the four metrics: The number of nulls for observation X must be equal to 0.
Prevent high variability in the standard deviation metrics.	The standard deviation of the columns used by the churn model.	The rules will trigger anomalies if the variability of the standard deviation is too high. It will compare new observations with the previously collected ones. The boundary can be set by trial and error. If you see a lot of false positives (anomalies that have no real impact on the model performance), you can readjust the boundary.

Table 5.1: Objectives of our use case

In the next section, you will see how these rules can be used in order to continuously validate the data at runtime.

Project – continuous validation of the data

Now that we have learned how to define the SLOs of our projects and how to transform these SLOs into rules, it is time to learn how to integrate these rules into a CI/CD process and how to implement an end-to-end data validation pipeline.

Concepts of CI/CD

For several years now, software development has adopted a set of best practices called CI/CD, which is aimed at eliminating the distance that exists between development and operations activities. This objective is mainly realized by forcing teams to automate the building, testing, and deployment phases of applications.

The acronym CI/CD stands for **continuous integration** (**CI**) and **continuous delivery** (**CD**), or in some cases, **continuous deployment** (**CD**). Before introducing the concept of continuous data validation, it is important to understand these concepts in detail.

In *Figure 5.2*, we can see a graphical representation of the main stages of these processes:

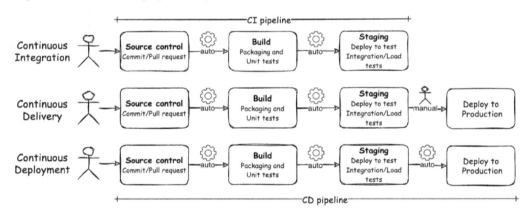

Figure 5.2 – Common CI/CD workflows

Continuous integration

CI is a development practice that requires developers to integrate their code changes in a shared repository using a version control system, such as Git, with very frequent releases in order to simplify the code merge phase with the modifications of the other contributors. The best practices indicate that the development of the code must be pushed and integrated at least once a day to avoid big differences with the evolution of the code base.

The following are the main common steps composing a classic CI process:

1. The developer modifies the code base on their local development environment.

2. The developer triggers the CI pipeline. Usually, it can be triggered in one of the following ways:

   - Automatically, pushing one or more commits to the shared repository

   - Automatically, creating a pull request to a target branch

   - By running the pipeline through a manual action

3. The first step of the pipeline is the build stage, where the code is compiled and then tested against a unit tests battery.

4. If all the unit tests pass successfully, the next step is to create an installation package, otherwise the CI pipeline stops with an error that is notified to the developers.

5. In case the CI pipeline has been triggered by a pull request, the code is merged to the target branch.

There are several advantages to the adoption of CI. For example, it helps to do the following:

- Promote automation, freeing developers from the burden of performing manual operations, especially for tests and builds

- Reduce the time needed to integrate the code between developers

- Promote small, incremental development, making possible the testing and availability of the most frequent version

- Identify anomalies in the code as soon as possible; in fact, a fundamental part of the CI process is that it automatically builds the application and executes the unit tests on the new code changes to immediately surface any errors

The adoption of the CI process has now become a standard in software development and, as we will see later, it also has a fundamental role in the continuous validation of our data.

Continuous delivery and continuous deployment

Continuous delivery is one of the most common software development practices and can be easily represented as an extension of CI. Indeed, it still comprehends all the automated steps that compose the CI pipeline, and expands upon CI by deploying all code changes to a testing environment and/ or a production environment after the build stage. These are the main further steps expected in a common CD process:

1. Automatically build a staging or testing environment and upload the installation package just built to make sure the software is working correctly.

2. Automatically execute the functional tests on the staging environment. For example, use dedicated tools that simulate the user behavior of clicking through the user interface.

3. Automatically execute load and performance tests on the staging environment.

4. After the package is created and all the tests pass with success, the new version of the application is potentially shippable. All that remains is to understand how to perform the deployment phase.

5. The deployment on the production environment is triggered by a manual action, usually by the delivery manager.

These further steps bring several advantages to the teams that adopt this practice:

- Automating the deployment reduces operational costs drastically. Indeed, the entire process is expected to be 100% automated except for the delivery of the production environment.

- Reduced time to market, it constantly provides an updated, reliable, and tested release of the software ready to be deployed and used by the users.

- It encourages a high frequency of deployment on production, anticipating the reception of feedback from users. The number of bugs being brought to the production tends to decrease over time.

The single difference between continuous deployment and continuous delivery is that for continuous delivery, every code change is built, tested, and then pushed to a non-production testing or staging environment, whereas continuous deployment also includes automated deployment in production. We can therefore consider continuous deployment as a sort of completion of continuous delivery. The main goals of this approach are to eliminate any manual actions and release into production with an even greater frequency.

Deploying the rules in a CI/CD pipeline

Now that we have acquired and consolidated the notions of CI/CD, we can understand how it is a fundamental part of data observability to monitor your data applications by constantly tracking your own statistics and inserting the rules within a CI/CD pipeline.

As we have already discussed in *Chapter 3: Data Observability Techniques*, there are different approaches to data monitoring. What we will see now is a concrete example of synchronous data monitoring, building a CI/CD pipeline for a Python data application.

Analyzing the data application

Let's start with the data application that we are going to instrument. It's called *Customers Enricher* and is a very simple Python data application that is responsible for retrieving all the customers' data and enriching it with other customer information, such as contacts and other business indicators.

Here is an overview of this application with its main components:

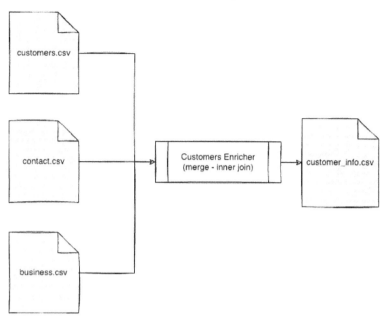

Figure 5.3 – Input and output of our data application

The application takes three CSV files as input:

- customers.csv: This file contains all the information strictly related to each customer, as shown in the following screenshot:

HTML Table Preview

age	job	marital	education	default	housing	loan	id
38	admin.	divorced	university.degree	no	yes	no	35704
33	services	single	basic.6y	unknown	yes	yes	16354
40	blue-collar	married	basic.4y	unknown	yes	no	10501

Figure 5.4 – Snapshot of the customers.csv file

- contacts.csv: This file reports all the information related to the contact that the company has had with the customer, as shown in the following screenshot:

HTML Table Preview

contact	month	day_of_week	campaign	pdays	previous_stat	poutcome	id
cellular	aug	mon		2	999	0	nonexistent	35704
telephone	may	wed		8	999	1	failure	16354
cellular	jul	wed		4	999	0	nonexistent	10501

Figure 5.5 – Snapshot of the contacts.csv file

- `business.csv`: This file contains several business indicators related to each customer, as shown in the following screenshot:

HTML Table Preview

emp_var_rate	cons_price_idx	cons_conf_idx	euribor3m	nr_employed	id
1.4	93.444	-36.1	4.963	5228.1	35704
-1.8	92.893	-46.2	1.281	5099.1	16354
1.4	93.918	-42.7	4.962	5228.1	10501

Figure 5.6 – Snapshot of the business.csv file

These three CSV files are then merged using the unique identifier of the customer. Then, the application produces and stores the output in a new file called `customers_info.csv` that contains one row per customer representing a full overview of the status of that customer.

Building the CI/CD pipeline

There are several tools we can use to implement a CI/CD pipeline, such as GitHub Actions, GitLab, Atlassian Bamboo, and Jenkins.

A CI/CD tool that provides all the steps needed to automate the phases of your CI/CD pipeline. You can create an automated process by defining a workflow – that is, a file containing all the instructions needed to trigger the execution of your pipeline – every time an event occurs in the source code repository of your application. A typical event is a commit pushed to the repository or a pull request being opened by a developer.

Your workflow can contain one or more jobs that can run in sequential order or in parallel. Each job runs inside its own virtual machine runner and contains one or more steps that either run a script defined by you or run an action – that is, another application – that you usually want to recycle in different workflows.

Our workflow contains three jobs. Here is the main structure:

1. Workflow > jobs

    I.   Build and tests

        i.   Set up Python

        ii.   Install dependencies

        iii. Lint

        iv. Execute unit tests

    II.   Staging

        i.   Set up Python

        ii.   Install dependencies

        iii. Execute staging

    III.   Production

        i.   Set up Python

        ii.   Install dependencies

        iii. Execute production

The following is a detailed diagram representing the same workflow:

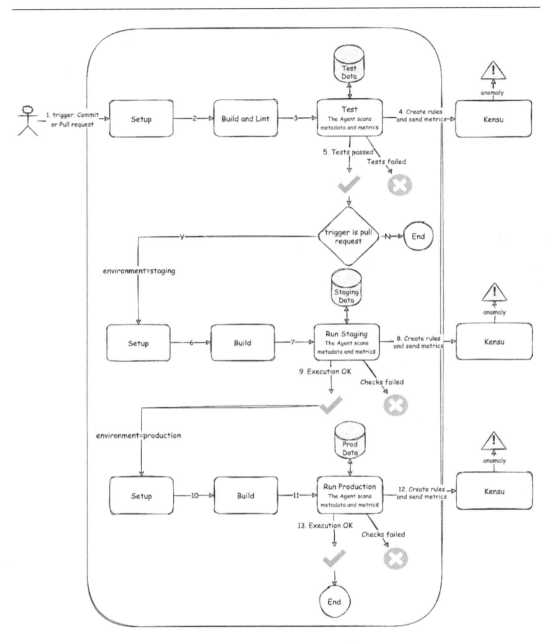

Figure 5.7 – GitHub Actions workflow

Summary

In this chapter, we learned how to define SLOs and how this can be done at different levels of abstraction, depending on the purpose, and analyzed the methods you can use to define your SLOs at the data source and project level.

Then, we learned how to turn our SLOs into actionable rules by defining and creating expectations that form the backbone of our rules.

By studying parts of the code, we have understood the different types of rules and their concrete implementation, as well as the concept of circuit-breakers.

In the last section of the chapter, we introduced the concept of continuous integration and continuous delivery to implement a positive and automated cycle of data validation.

Part 3:
How to adopt
Data Observability in
your organization

This section advances into the practical implementation and cultural integration of data observability. It begins with an exploration of root cause analysis, showing how data observability can streamline and automate anomaly detection and troubleshooting, including methods for data incident management and examples. Next, the focus shifts to optimizing data pipelines, highlighting how observability aids in managing aspects like cost efficiency and risk. The final part discusses the introduction and cultural adaptation of data observability in teams. This includes an examination of various data team structures, their integration into different organizational types, and strategies to measure the success of data observability initiatives.

This part has the following chapters:

- *Chapter 6, Root Cause Analysis*
- *Chapter 7, Optimizing Data Pipelines*
- *Chapter 8, Organizing Data Teams and Measuring the Success of Data Observability*

6

Root Cause Analysis

Creating rules and expectations is one thing, as it allows you to detect any issues in your data, but troubleshooting is another.

An observed system should give you many clues and means to check the origin of the error, which will lead to efficient data issue resolution.

In a company, resources are key. The team's time should be dedicated to value creation, not maintenance or troubleshooting under pressure. You need to know how to use the resources efficiently to avoid wasting time and, ultimately, money.

The best way to keep these costs under control is to evaluate them using **key performance indicators (KPIs)**. Some interesting team or project metrics that you may like to follow include the mean time to detect and the mean time to resolve. The former designs the period between the incident's occurrence and its detection, while the latter describes the amount of time spent resolving the issue. The goal of the head of data, and all data engineering teams, is to eliminate breaches of the agreed-upon KPIs as fairly as possible. In this chapter, we will see how efficient data observability is necessary for that purpose.

In this chapter, we'll activate the metrics and rules we covered in the previous chapters to present a method you can follow in case an incident has been detected and a case where this method can be applied.

This chapter focuses on how to address anomaly detection so that we can identify problematic data sources and applications involved in our data pipelines.

In this chapter, we'll cover the following topics:

- Data incident management
- Anomaly detection

Let's start by seeing what happens when an issue is detected.

Data incident management

When an issue is detected, the team's productivity can be affected as resources mobilized on issue resolution cannot be used to create value with new projects. Therefore, to avoid working under pressure and troubleshooting the issue in an unsustainable way, the method we propose is as follows:

1. Detect the issue.
2. Evaluate its impact.
3. Find the root cause.
4. Troubleshoot.
5. Avoid future similar issues.

Thanks to observability, each step will be supported by logs, metrics, and traces that you can use to reduce the time the team spends on resolving issues.

Let's explore each step in more detail.

Detecting the issue

An issue can be detected by several means. Let's say that, before you read this book, the majority of the issues are reported to you. One of your data providers or internal customers could come to you to signal that *something is fishy* in the data you are consuming or, even worse, that the data you are producing leads another team to wrong decisions.

Of course, this is a situation we want to avoid. The rules we introduced and discussed in the previous chapter are our great allies here. Thanks to the rules and the corresponding checks, the issue will be detected once the pipeline has run or even during its execution. On top of that, they can even prevent the issue from being replicated and propagated to the chain of data applications and data sources belonging to the same lineage.

Without efficient issue detection, latent errors may only be detected when the data is used by the consumers, leading to disastrous effects. Besides this, a person can be fully dedicated to assessing the quality of the data after each run, creating some delay between the execution and the data issue detection.

First, we must distinguish two notions that are involved when detecting issues: the event and its notification.

On one hand, the event is the issue detection itself as it detects an anomaly in the process. On the other hand, the notification is how the event will be shared with the stakeholders.

The event should contain as much information as possible. You will start an investigation; you need to be fully aware of the issue you will treat. The context component of data observability is highly relevant here and needs to answer the following questions:

- Which data source is involved?
- In which application did we discover the issue?
- When did the issue occur?
- In which environment and project did it happen?

The more details you have, the clearer the view you will have on the issue.

This notification is important for addressing the issue as soon as possible and contributes to the reduction of the mean time to detect. Data observability metrics should not be manually checked after each run. Instead, informing the right people once the issue occurs is a success criterion to establish confidence in the data. Thanks to the observability context, each issue can be linked to the right stakeholders, to everyone who has a real interest in being notified about the issue. The team responsible for the data source, the so-called producer, needs to know if an issue was detected so that they can take action and fix it. The consumers of the data may also need to be cautioned that the data they will use for another application or decision-making is faulty.

Once the issue has been detected, it is time to assess its scope. This can be done by performing efficient impact analysis.

Impact analysis

In a company, resources are always limited. Jumping on the slightest problem may make you lose time and allocate your efforts to the wrong assignments, thereby impacting the overall productivity of the team. To prioritize tasks and problem resolution, it is common to start with impact analysis before trying to fix any problem.

Impact analysis is part of the definition of the problem – it consists of listing all the stakeholders, applications, projects, and data sources that are affected by the detected issue, and comprehends the evaluation of the criticality of the situation based on the impact it may have on other teams' work and the main KPIs of your company. Thanks to this analysis, your team will be able to make informed decisions while they're prioritizing the intervention they will need to bug-fix the data. For instance, an issue that impacts a $10 million process will likely be more urgent than a relatively small $10k issue.

Without observability, deep knowledge of the architecture diagram is needed to assess how the error could affect the data. Dealing with all the projects of the data environment can also lead to errors and the risk of overlooking a stakeholder is pretty high. Also, there is a balance to maintain between putting all projects on hold during the investigation and letting them run with the risk of replicating issues.

This is where observability unleashes its added value. The collected metrics and their context will help define what is impacted by the issue.

This downstream lineage can be used to list all the associated processes and data sources. Other elements can also be used, such as the frequency of the runs of the dependent applications. If the interval period between two runs is large, the urgency of solving the issue can diminish. However, if a dependent data source produces a report that's used in a daily sales meeting, you may need to solve it faster.

Once the issue has been clearly defined and its impact has been evaluated, you can mobilize the producer team to solve the issue. The first step will be to define the cause of the issue. This can be done using root cause analysis.

Root cause analysis

The root cause of a data issue is the source issue that produces a series of issues along the data pipeline. Root cause analysis is an incremental process where the analyst asks themselves *why* each time an anomaly is detected.

When an issue is detected, the cause of the issue can be another issue itself. For instance, an ingestion process may receive outdated data and copy it to the master data. If the issue is detected at the master data level, this means that the input of the ingestion application is still faulty, and its creation lineage should be analyzed to find the cause. This will help you identify the initial cause of the master data issue. This is explained in *Figure 6.1*:

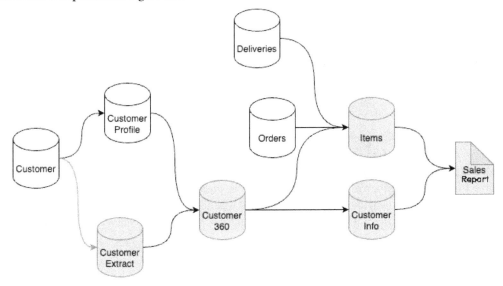

Figure 6.1 – Root cause analysis

As shown in *Figure 6.1*, a data issue has been detected in the final report. This issue can be directly traced back to the input databases. However, this issue was already present in the source feeding those databases.

Data observability, through the observability metrics it gathers, helps collect clues and evidence about the problem to better prepare its resolution.

Solving any intermediary step between the root cause and the detected issue will be useless. It is like taking a painkiller – the feeling of pain is soothed but will come back so long as the source of pain is not treated. Finding the root cause of the issue will also point out all dependent issues.

Here are a couple of different types of root causes you can usually find:

- **An issue in the data application**: These issues are internal to the company and can be fixed by the producers. Here are some examples:

  - An application running out of allocated memory so that the data could not have been produced on time or could have been produced partially

  - A change in the code is removing items from a data source

  - A missing application, decommissioned without consideration of its legacy usage

- **An issue in the data source**: These issues can be trickier and non-resolvable as they may not depend on an internal process. Here is an example:

  - When the root cause leads to a data source containing missing values, that was bought from another company, the initial cause in your filesystem is found. However, this is not identified as the root cause as another issue led to the production of this faulty data source.

Thanks to data observability, using logs, metrics, and traces, we can find the root cause of an issue without accessing the data or replaying applications.

The method we propose is as follows:

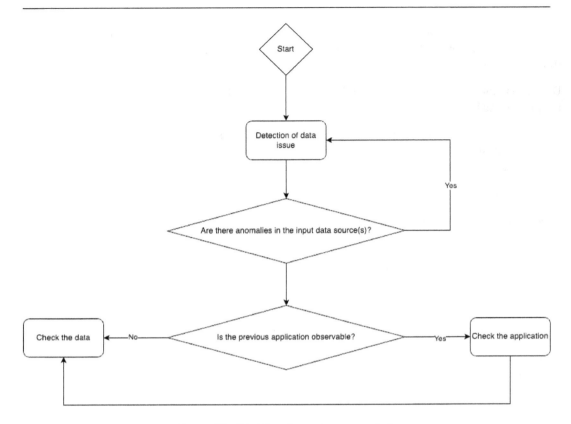

Figure 6.2 – Workflow for root cause analysis

When a data issue is detected, we need to iterate in each application that creates the data to see what input data source could have caused the issue. The first question we must ask is are there anomalies in the input data sources? This will result in the detection of other issues, but it may still not be the root cause.

Once the input data source of an application meets all the quality standards and it is no longer possible to attribute the fault to the input data source, it likely means that an issue happened inside the application itself. However, if you come to the end of the investigation and figure out that the original input data is wrong, it may signal another issue upstream, outside of your scope of observability.

Without data observability, detecting issues would be time-consuming. First, you need proper access to the data used by the application, if and only if you know what data is used. The same applies to the applications themselves: checking the code base may be restricted to some users. You may consider handing over the investigations to the owner of the source application.

When the root cause of the issue has been detected, it becomes possible and easier to fix. Let's see how observability can help with troubleshooting.

Troubleshooting

If the issue is coming from the data you have received from an external provider, there is not much you can do.

However, if you need to fix one of your company jobs, the troubleshooting process is made easy by identifying the contextual information around the root cause issue. Elements such as the code owner, code location, and code version are essential to fix the issue at the source and are also essential information that you can share with your external provider to support its root cause analysis. Observability avoids the tedious process of comparing the code bases to find the one creating the issue.

Finally, once the issue has been solved, you need to prevent it from reoccurring.

Preventing further issues

Once the issue has been solved, preventive actions can be taken to avoid similar root causes from happening again in the future. This will also increase the speed of future root cause analyses by helping you discard known potential causes of issues. This can be done by adding new checks or refining the existing ones. Without observability, similar issues may lead to the same tedious manual approach to troubleshooting.

Applying the method – a practical example

Let's go back to our example from *Chapter 2*. In that example, you were in charge of a small notebook that ingests data into a SQL database. We identified several objectives linked to the outcome of that notebook. Here, we'll see how data observability at scale could help solve the detected errors. When observability is applied to all the company projects, you can create an application lineage map, like this:

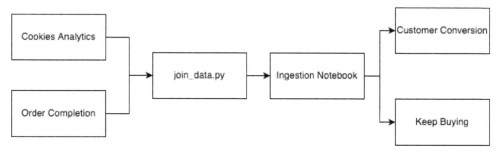

Figure 6.3 – Application lineage example

Let's discuss the structure of the projects. You are responsible for the ingestion notebook application. However, as you can see, this application relies on several other applications.

The data you obtained in the S3 bucket was created by another application, `join_data`, which itself was fed by different processes. The full pipeline can be described as follows:

1. First, an application named *Cookies Analytics* collects the cookies from the website via an API and creates a list consisting of the email, the URL of the visited page, the order ID, and a timestamp. This is summarized in a table called `cookies_info`.

2. Simultaneously, an application named *Order Completion* takes care of recording what the user puts in their basket and checks whether the order has been paid for or not, creating a dataset containing the ID of the order and the amount of the order.

3. Finally, an application called `join_data` joins the two created datasets and computes KPIs. It will compute the duration, as well as the total number of pages visited by the user. This final CSV file is put in an S3 bucket and is the one you read when performing ingestion. It is the input data of your part of the pipeline. The output is the `order` table.

Downstream, as a reminder, the data source is used for two marketing campaigns: *Keep Buying* and *Customer Conversion*.

Detecting the issue

SLOs have been defined at the producer level to ensure the data that was produced maintains high-quality standards that have been agreed upon with consumers. The first rules you defined are as follows:

*   **Data must be complete**: It needs to contain the `order_id`, `email`, `duration`, `total_basket`, `page_visited`, and `has_confirmed` columns

*   **Data must be fresh**: The data that's sent to the data lake must contain the events of the past month

Let's focus on the issue we had in January. In that example, the data was incomplete because of a freshness issue. Data for several days was wiped out for some reason.

To detect the issue, you could have had a rule that checked whether for each day, you had at least one row in the dataset. The number of days without any rows can be an indicator. When the application runs on January 31, an event is automatically created: if the number of incomplete days is greater than 0, there is a data error.

The stakeholders of the data source must be immediately informed of the data issue. You, as a producer, need to take immediate action to gauge the impact of the issue, while consumers need to be made aware so that they can avoid using the produced dataset or at least know that there are limitations and known bugs.

Impact analysis

Once the issue has been detected, the first step is to find out who and what it is impacting or could be impacting soon. To do so, you can use downstream lineage. As shown in *Figure 6.3*, the direct impact is on the two marketing projects – *Customer Conversion* and *Keep buying*. However, it is interesting to note that those projects are also the foundations of other data applications inside the company, so more teams will be potentially impacted by a data issue cascade.

Now, the team has to evaluate the criticality of this issue. First, from a business point of view, the reports that are created by the application are used in many projects and could lead to misleading decisions as the data is incomplete.

Second, technically speaking, you have time to tackle this issue before the marketing team needs the data. Thanks to the contextual information about the execution that observability gives, the frequency of the run of the dependent projects can be inferred. As shown in *Figure 6.4*, no run of the project is due before one week's time. The data source that's influenced by the error is used by two projects on the 7th and the 15th of each month. The issue was detected on January 31, which buys you some time to solve the issue:

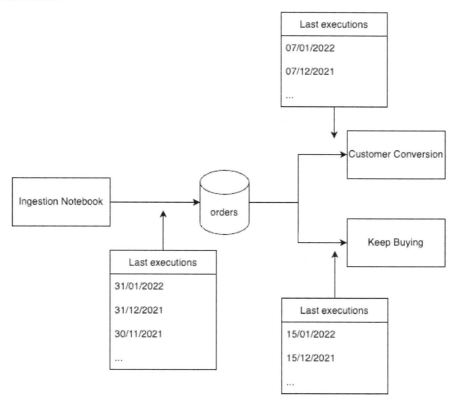

Figure 6.4 – Impact analysis example

If the team considers it a priority, you will need to solve the issue. To do so, let's explore the root cause.

Root cause analysis

Finding the root cause of the issue is an iterative process where we try to find anomalies in the applications that were used to create the producer's data source.

Using our method, we have several applications and data sources to check. First, inside your proper application, is the input already presenting the freshness anomaly? Let's focus on the metrics of the ingestion notebook shown in *Figure 6.5*:

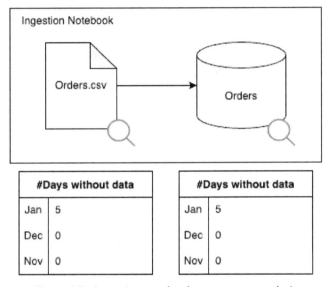

Figure 6.5 – Ingestion notebook – root cause analysis

As the metrics indicate, the issue is already present in the input source. The five days without data that were detected during January executions are already missing in the input data. This means that the root cause is elsewhere upstream of the application.

The next step consists of checking the path the data has used to feed your application. Thanks to the lineage, you can see that join_data was used to create the input source of the ingestion application. Now, you need to check whether there are any anomalies in this application. Again, looking at the observability metrics of join_data will help. This can be seen in *Figure 6.6*:

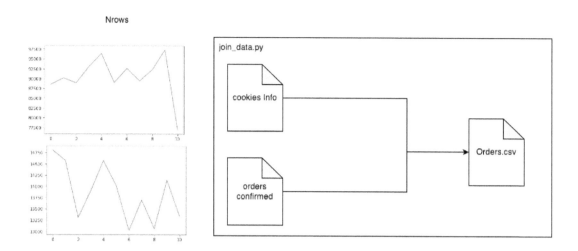

Figure 6.6 – join_data.py – root cause analysis

Here, you can see the two data sources that were used to create the joined `Orders.csv` file. This time, you don't have any visibility about the number of missing days. This metric is not part of the observability metrics that were collected in the `join_data.py` application.

You will need to rely on other metrics to find the origin of the error in this application. Interesting fact – the number of rows of the *Cookies Info* data source is significantly lower this month compared to other executions. We can plot the last 10 executions; upon doing so, we can easily realize that while the number of orders is stable at around 14,500, the number of visited pages decreased from ~90,000 to ~75,000. This is a drop of more or less 15%, which corresponds to the 5 days of data missing in the `Orders.csv` table. This is an anomaly. However, this is not the root cause in our case; further analyses are possible.

Now, let's see how *Cookies Info* was created. The application that's creating *Cookies Info* uses information retrieved from an API. Checking the number of rows of the input and the output will help you define the root cause:

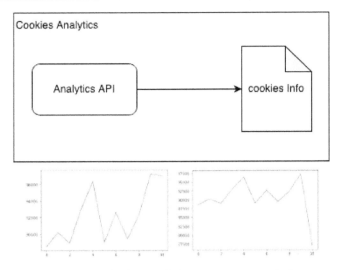

Figure 6.7 – Cookies analytics – root cause analysis

In the input data source, we cannot see the same drop as in the output. Comparing both time series on a single graph will help us evaluate the magnitude of the issue:

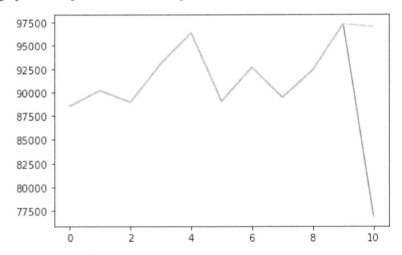

Figure 6.8 – Comparison of nrows time series

We can see that while the number of data items sent by the API remained high, the data that was written to the output file was lower in January than the amount of read data.

Congratulations! You have found where the root cause of the issue is nested. Observability also provides you with the context of the metric. It is time to troubleshoot it.

Troubleshoot

Now that you know where the issue is located, you can start investigating what happened inside the application to create the data issue. Thanks to the observability metrics, you can see that the newest version of the data application is the culprit:

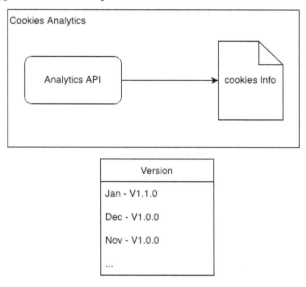

Figure 6.9 – Cookies analytics – latest executed versions

Information such as the code base's location and version can help you retrieve the right branch to start the resolution process.

With this kind of information, you can dig into the code and compare the versions. Interestingly, you will see that a new version of the code was released in January, taking into account some new changes in the API. However, the new API was released on the 23rd while the current code was put into production on the 18th. To solve this issue, the data source must be populated with the missing data, which is easy in this case. The application must be rerun with the correct version of the code for the missing days.

Let's see how we can leverage our discoveries to prevent future errors.

Preventing future issues

To avoid this issue in the future or to dismiss similar causes once another issue has been detected, you can add several indicators.

First, following the root cause analysis process, you can see that some metrics could have helped you see whether the issue was replicated. For instance, you could have retrieved the number of data rows per day or the missing days per month for the `join_data.py` application.

Second, you could have added some variability rule to the number of data rows you write in the *Cookies Analytics* process. If the indicator varies a lot compared to the previous execution, an alert could be issued.

Finally, you could add a circuit breaker inside your application, raising an exception if the number of rows in the input data source differs from the output. This may delay the data's availability to the stakeholders but it will prevent them from using the wrong data.

Advantages of the data observability method

In this example, we saw how data observability can help in troubleshooting a data issue. It all started with contextual logs and traces gathered at the time of the run. Thanks to this, issues can be detected and dataflows can be understood. The troubleshooting process is made simpler and allows you to control quality in the context of all the data sources used for a certain purpose.

Without observability, the same issue discovery process would have taken weeks or months in a large company. There, you would have needed the correct rights to query the data sources outside the scope of your work. In this example, being responsible for the output of the ingestion notebook does not guarantee that you have access to the data sources that create your inputs. Requesting suitable rights may take a considerable time, slowing down the process. With observability, the system has regularly reported about its state. Thanks to this, no data access was needed during the troubleshooting process.

In reality, when performing this at scale, a lot of metrics can be used and can pollute the root cause analysis. To find out what metrics are interesting, you must follow the anomaly detection process, which reduces the set of lineages you have to explore. This is what we will explore in the next section.

Anomaly detection

In this section, you'll learn how an anomaly can be detected based on the metrics you have gathered and can be the basis for rules starting from the SLIs.

Anomalies can be inferred from the following:

- Simple indicator deterministic cases
- Multiple indicators deterministic cases
- Time series analysis

Let's dig into these categories to explain how the observability metrics can be leveraged to find out the root cause and, *in fine*, new points of attention and rules for your data sources.

Simple indicator deterministic cases

Anomalies can be detected by adding a series of basic checks to the rules based on the type of metrics you gather, as well as the business logic.

By handling missing values effectively, organizations can prevent potential misinterpretations or errors in data analysis. For example, if the data producer or consumer expects no missing values in the data source, a deterministic rule addressing the number of missing values can be established. In cases where the collected metric represents a counter for missing values, we could constrain this value to 0.

This approach to anomaly detection can be customized for different metrics. For example, ages would typically fall within the 0-120 range, while a company's turnover is expected to remain stable annually or grow according to set objectives. Similarly, there could be a referential data source that you would expect to remain consistent month to month. Monitoring these metrics can help you identify discrepancies that may indicate issues in data collection or processing.

Anomalies can also manifest in other aspects of data observability, such as an unexpected alteration in a supposedly stable schema, or a sudden change in the frequency of an application's data updates. By keeping an eye on these additional factors, data teams can ensure that their data processes remain robust and reliable.

The following table summarizes the deterministic rules that can be inferred from the indicators:

Indicator(s)	Example of a Deterministic Check
Schema	The schema is stable over time
Null rows	Null rows cannot be more than 0
Lineage	The lineage is creating the same number of columns over time
Application runtime	The period between two runs is supposed to be constant

We can also extend our analysis by comparing several indicators. We'll look at this in the next section.

Multiple indicators deterministic cases

The indicators can also be combined to give you other insights about the data. If instead of giving an absolute range, you want to compare the missing values based on the percentage of rows presenting an anomaly, you will need to compare the metric counting the missing values with the one counting the number of rows.

Combining multiple indicator rules can also bridge the gap between application observability and data metrics. For example, by ensuring that an application with a bug fix version returns consistent results as before the update, you can maintain the integrity of your data and the reliability of your applications.

Time series analysis

When executions start accumulating, patterns can be found in the data observability metrics that you gather. These patterns are extremely powerful in detecting anomalies over time. However, the real insights may take time to be computed as more data points have to be collected. Although deriving these insights may require time and a sufficient number of data points, gathering the metrics at runtime for each execution can significantly reduce the time needed to address issues when they arise as you won't need to reconstruct the historical data metrics.

Anomalies can be detected if the expected metric is different from the collected one. To define what the metric should be, you can apply various machine learning methods. Some of those methods rely on a couple of concepts: trend and seasonality.

Trend

The **trend** explains how the value is expected to increase or decrease over time. Some parts of the time series can have an increasing trend, while others can have a decreasing trend. The trend is not necessarily linear and is not influenced by the season.

For instance, if you expect the sales to grow over time, the trend will likely be increasing.

Seasonality

Seasonality refers to the influence the season will have on the time series. The season is a fixed period such as a week or a month and predicts local minimums or maximums during this period. The time of the year or the day of the week affects the expected value. For instance, a backpack store can expect a peak in loyalty card registration each year during the *BackToSchool* action or the Christmas sales.

By combining trend and seasonality, you can easily approximate the expected value for a certain execution. If the observed value deviates a lot from what was expected, this constitutes an anomaly. Further analyses, such as insights from the business, can help you confirm or reject the anomaly.

Methods that combine these notions include the Holt-Winters seasonal method, the exponential smoothing method, and advanced models such as (S)ARIMA.

When implementing anomaly detection using machine learning methods, it is important to continuously update and refine the models as new data points are collected. This ensures that the models remain accurate and effective in detecting anomalies, even as the underlying data and business environment change.

Case study

In this section, we will show you some examples of anomaly detection algorithms that can be performed on indicators.

You can follow the results in the notebook associated with this chapter in this book's GitHub repository, called `Detect_Anomalies.ipynb`.

We're focusing on the time series data of metrics, or **service-level indicators** (**SLIs**), that have been observed over a certain period. The dataset, named `observability_metrics.csv`, is quite insightful. It presents a list of metrics for each execution of an application, specifically analyzing the `orders` dataset at runtime. This application runs daily, and each day's metrics are neatly organized into a row in the dataset.

Let's look at the collected metrics, which form the columns of our dataset:

- `nrows`: This represents the total number of rows in the `orders` dataset
- `orders.quantity.mean`: Here, we calculate the average quantity of goods across all orders
- `orders.price.mean`: This metric reflects the average price across all orders
- `buyer.cookie.duration`: A key metric indicating the average time a buyer spends on the website

By exploring these metrics, we aim to uncover any anomalies that could indicate important insights or potential issues in the dataset.

Anomalies are not outliers in the dataset. This distinction is crucial. Our focus is on identifying anomalies within the observability metrics computed from the application's dataset, not on pinpointing outliers within the dataset itself. This means we are scrutinizing how these metrics deviate from their expected patterns or norms, which could indicate potential issues in a given context (here, the application).

However, it's worth noting that outliers in the dataset can still be significant. While they are not our primary concern in this context, these outliers could potentially be tracked as SLIs if they provide valuable insights into the application's behavior.

Simple indicator deterministic cases

In this section, we'll shift our analytical focus to the `buyer.cookie.duration` column of our dataset and employ univariate anomaly detection techniques. This column, representing the average time a buyer spends on the website, is a critical metric for understanding user engagement and website performance. Detecting anomalies in this metric can provide insights into user behavior, website usability issues, or potential data recording errors.

Univariate anomaly detection involves analyzing a single variable to identify data points that significantly deviate from the norm. These deviations are considered anomalies or outliers. The approach we adopt here is statistical, focusing on the distribution of `buyer.cookie.duration` values over the observed period.

The first step in our analysis involves understanding the typical behavior of the `buyer.cookie.duration` metric. We calculate statistical measures such as the mean and standard deviation to establish a baseline of normal behavior. With this baseline, we then apply a commonly used rule for anomaly detection – identifying data points that lie beyond a certain threshold from the mean, typically more than two or three standard deviations away. Let's have a look at the time series shown in *Figure 6.10*:

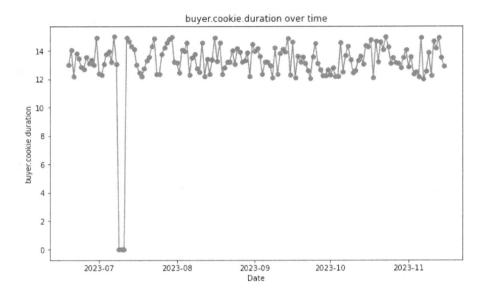

Figure 6.10 – Evolution of the buyer.cookie.duration metric over time

From this chart, it is obvious that we have a drop of 3 days, when the duration became 0, indicating a potential issue in collecting the website cookies.

To check this programmatically, we can look at the numbers deviating too much from the mean. In this section, we will use the **Python Outlier Detection (PyOD)** library and the **median absolute deviation (MAD)** method. PyOD is a comprehensive toolkit for identifying outliers in multivariate data, but it can also be effectively used for univariate analysis. MAD is a robust measure of variability that is less sensitive to outliers than standard deviation.

Here's a concise overview of how it works:

1. **Calculate the median**: Determine the central value of the data.

2. **Compute absolute deviations**: Find the absolute differences between each data point and the median.

3. **Determine MAD**: Calculate the median of these absolute deviations.

4. For each data point, the anomaly score quantifies how much the point deviates from what is considered normal for the dataset. In the case of MAD, this involves how far a data point is from the median in terms of median absolute deviations.

This is particularly useful in scenarios where data might not follow a normal distribution. In *Figure 6.11*, we can see the MAD score for our data points. A MAD score above 3 is generally used to define a point as being an anomaly:

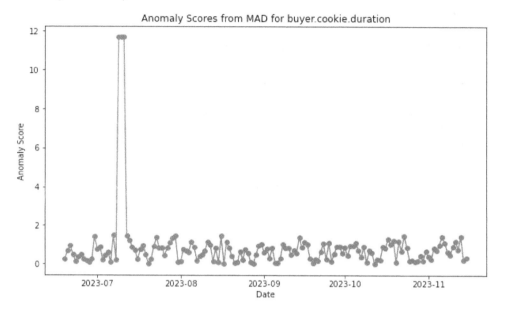

Figure 6.11 – Anomaly score from the MAD technique

From this chart, it is made clear that the 3 days with 0 as a duration are outliers.

Knowing the usual behavior of the data, you can now better monitor the dataset by adding additional rules, such as a rule that will trigger an alert if the new data point is far from the median of the historical values.

Multiple indicators deterministic cases

Now, let's explore the correlation between two key metrics: `orders.price.mean` and `orders.quantity.mean`. Our approach will dissect this relationship week by week, offering a dynamic view of how these two variables interact over time. It is an example of how two indicators can be used to build a new indicator that can be tracked and monitored.

Our primary aim is to understand how the average price of orders (`orders.price.mean`) and the average quantity of goods in all orders (`orders.quantity.mean`) are related. Do higher prices correspond to greater quantities, or is there an inverse relationship? Or, perhaps, the relationship varies over time.

We must divide our dataset into weekly segments. For each week, we must calculate the Pearson correlation coefficient between the two metrics. This coefficient, ranging from -1 to 1, will indicate the strength and direction of the relationship:

- A value close to 1 suggests a strong positive correlation (as one metric increases, so does the other)

- A value close to -1 indicates a strong negative correlation (as one metric increases, the other decreases)

- A value near 0 implies little to no linear correlation

This analysis is essential in understanding consumer behavior and market dynamics. For instance, a consistently positive correlation might suggest that customers are willing to purchase more at higher prices, possibly indicating a perceived value or quality. Alternatively, a negative correlation could imply price sensitivity among customers.

Figure 6.12 describes the evolution of the correlation:

Figure 6.12 – Weekly correlation between price and quantity

The consistent presence of correlation above the 0.7 mark across the weeks suggests a robust and persistent positive relationship between the two variables throughout the observed period. However, depending on the business outcome, the sudden drop to 0.6 may indicate an anomaly and an underlying issue in the data.

A new rule can be derived from those two indicators (`orders.price.mean` and `orders.quantity.mean`), monitoring the correlation between both metrics.

Time series analysis

In this section, we'll focus on the `nrows` component within the initial 6-week period of our dataset. *Figure 6.13* shows the evolution over time and adds the day of the week. We can immediately notice that weekends have fewer orders (as a reminder, one row in the original dataset is one order), while there is a surge on Mondays. However, we can see that there is an anomaly on the second Thursday, where the number of orders skyrocketed:

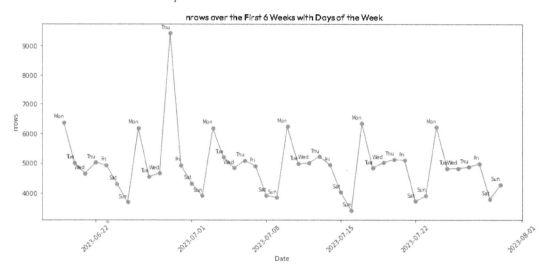

Figure 6.13 – Time series of the nrows metric

A focused analysis, which can be achieved through time series decomposition, splits the data into three key elements: trend, seasonality, and residuals. This approach is essential for revealing the inherent structures and irregularities in our dataset. *Figure 6.14* shows a seasonal decomposition of the time series:

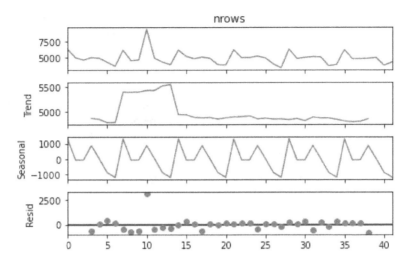

Figure 6.14 – Seasonal decomposition of the nrows metric

Let's break down and analyze all the components.

Trend analysis

The trend component offers a macro view of the dataset, showcasing the overarching movement over 6 weeks. Our analysis revealed a subtle yet discernible trend, possibly indicative of the evolving market dynamics or operational shifts during this period. This gradual change in the nrows value could suggest a steady increase or decrease in transaction volume over time, shedding light on the broader business trajectory.

Seasonality analysis

Seasonality, on the other hand, brings to light the repetitive, predictable patterns that occur at regular intervals. In our dataset, these patterns reflect the weekly cycles of business operations. For instance, certain days of the week might consistently show higher or lower transaction volumes, indicating routine customer behavior or operational rhythms. This insight is invaluable for planning and forecasting, allowing us to anticipate regular fluctuations in transaction volume.

Residuals and anomaly detection

Finally, the residuals – data remaining after accounting for trend and seasonality – are where we spot anomalies. In our case, the most notable outlier appears on June 29, 2023, where the residual value of about 3119.58 starkly deviates from the norm. This suggests an unusual spike in transactions, possibly due to a special event or an error in data recording, warranting further exploration. The same techniques as the ones used in univariate analysis (MAD or standard deviation) can also be used on the residuals to confirm the anomalous character of the data point, and similar rules can be created.

Through this detailed analysis, time series decomposition has proven invaluable in dissecting and understanding the dataset's complexities. It has enabled us to not only comprehend the regular patterns and long-term trends in our data but also to pinpoint and investigate the anomalies that could lead to deeper insights or reveal underlying issues.

Now, let's summarize what we've learned in this chapter.

Summary

In this chapter, we saw the real value of data observability for data engineers, where they troubleshoot or even firefight issues in their day-to-day jobs. Days or weeks of tedious manual checks can be avoided by adding proactive and at-the-source observability. The observability metrics that are collected by applications moving, reading, and transforming the data are great assets for performing analyses in case any issues occur.

Furthermore, we have seen that the more observable the system is, the easier it is to evaluate the impact of any issue, allowing the team to work efficiently on what requires the most attention. The in-context collected metrics allow us to easily overview the content of the data through the lineage so that we can correctly identify the faulty application or data and fix it faster.

This is only one of the main advantages of implementing data observability. In the next chapter, we will explore how data observability can be used to optimize pipelines.

7
Optimizing Data Pipelines

The importance of data in companies has significantly increased the investments in data platforms by companies. Over time, this has increased companies' priority of being aware of what their data pipelines do and how they do it and therefore monitoring not only the quality of the outcomes but also the state of health of the pipelines. At the same time, they are also monitoring the usage of the resources and tracking the associated costs.

In this chapter, we will understand how data observability offers us a way to make the governance of our data pipelines scalable and sustainable. First, we will focus on understanding the key data pipelines, their main components, and the types of data pipelines, as well as their characteristics. Then, we will learn how data observability and, in particular, data lineage can be used to manage several aspects of the data pipeline life cycle, such as the costs and the risks.

In this chapter, we'll cover the following topics:

- Concepts of data pipelines and data architecture
- Rationalizing the costs

Concepts of data pipelines and data architecture

We rarely think about how water reaches the taps of our homes. After all, we are end users who pay and use the service with certain expectations and have little visibility and interest in what concerns the transport and management of drinking water.

But this is a good moment to stop for a few seconds to understand this process – it is a process that has many similarities with data pipelines.

What is a data pipeline?

To better understand what a data pipeline is, we can compare it to the components that carry water from the basins to our homes:

- There is a basin of water to draw from (the data sources)
- Various mechanisms are needed to recover, purify, and transport water (the data applications)
- The water reaches the taps of our houses (the data destination)

At this stage, let's define what a data pipeline is. It is the flow of data that starts from one or several places where data is stored. The data passes through several applications, after which a series of data processing steps is dedicated to reading and transforming the data. Each step delivers the output data that is used as input by the next step. The pipeline ends at the target data sources – that is, the destination.

Yes, it's very similar to the flow that's needed to transport water. The main components of each data pipeline are as follows:

- The data sources (the basin of water)
- The data processing step(s)
- The data destination, often called the data sink

This is depicted in the following diagram:

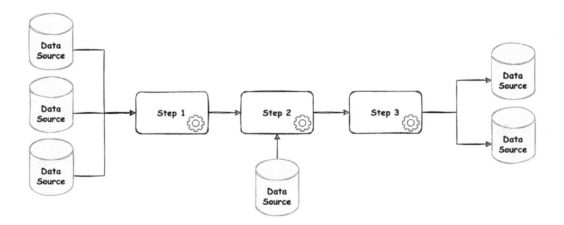

Figure 7.1 – A possible data pipeline structure

But what do these steps usually do? And what is the scope of each step? Well, there are a lot of different goals a data pipeline step could target. Let's look at some of these steps to better understand what a given pipeline can be composed of:

- **Data extraction or acquisition**: This involves gathering data from various sources and storing it in a central, sometimes temporary, location

- **Data transformation**: This is an essential step in the data preparation and analysis process that typically involves converting data from one format or structure into another format or structure by applying various algorithms and techniques to manipulate, filter, group, aggregate, enrich, and extract relevant information from the data

- **Data cleansing**: This involves identifying and removing missing, inconsistent, or irrelevant data, as well as detecting duplicate data or formatting the data so that it matches the format expected by the further steps

- **Data analysis**: This involves using statistical techniques or machine learning algorithms to identify insights, patterns, and trends from the data

- **Data reporting**: This involves summarizing the findings and presenting them clearly and concisely to stakeholders

- **Data visualization**: This is a particular type of data reporting where the data is presented in a visual format, such as graphs or charts, to make it easier to understand

- **Data sharing and distribution**: This involves making the results of the analysis available to stakeholders and decision-makers

Defining the types of data pipelines

Now that we finally know what a data pipeline is, let's understand the different types of data pipelines we have and how they are classified in a few different ways.

Classification based on the architecture of the data flow

Based on the architectural choices of the data flow, the data pipelines can be **Extract, Transform, and Load (ETL)** or **Extract, Load, and Transform (ELT)**, depending on whether the transformation step happens before or after the data is loaded into its destination.

It's also important to specify that, often, the concepts of ETL and ELT are confused with the term data pipeline. Let's understand what the differences are.

ETL consists of three phases:

1. **Extraction**: The first step is to move data from different data sources – for example, two tables belonging to two different database instances – to a single temporary common place.

2. **Transformation**: You transform the data gathered in the previous step to make it compliant with the expected result – for example, you can fill the empty fields using a default value.

3. **Load**: This is the last phase and is where the outcome of the transformation is loaded into the target destination – for example, a data warehouse.

This is depicted in the following diagram:

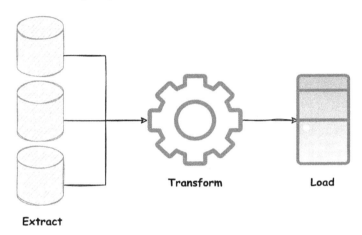

Figure 7.2 – ETL data flow

The ETL approach is typically used when you must use data from different data sources. In this case, since data is not in the same place, you need to extract and move it to a common place.

If you have to transform a huge amount of data, when possible, ELT is a more scalable and powerful approach. It comprises three phases:

1. **Extraction**: The first step is identical to the first step of ETL – it moves data from different data sources to a single common place.

2. **Load**: This step is a big difference between the two approaches: with ELT, the extracted data is not stored in a temporary place; instead, it could be immediately uploaded to the server where it will ultimately be used, but it could also be staged in an S3 bucket first and retained as part of data lake architecture, for example. This reduces the time between extraction and delivery.

3. **Transformation**: The database where the data is moved is then used to apply the data transformation that's been requested: .

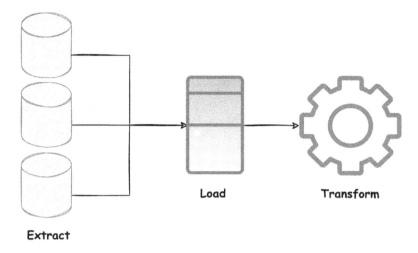

Figure 7.3 – ELT data flow

So, we can consider ETL and ELT as particular types of data pipelines, though in other cases, they are a particular type of subset of a data pipeline. In other words, often, multiple ETL or ELT processes make up a single data pipeline.

Classification based on the type of processing

Based on the type of processing that's performed, the data pipelines can be either batch or streaming pipelines, depending on whether they process data in batches or in real time.

The batch data pipeline processes large amounts of data together in a single batch or group, rather than individually in real time. This typically happens at regular intervals or on a schedule.

The batch approach is commonly used for offline data processing, such as ETL processes, data warehousing, and data lake management, and uses specialized software and hardware tools to process the data systematically and efficiently. The resulting output is typically used for analysis, reporting, or other business purposes, where large amounts of data need to be processed quickly and accurately.

The second type of data processing include a real-time data pipeline. This type of data pipeline processes data in real time, as it is generated or received. It is commonly used for online data processing, such as stream processing, event processing, and data analytics.

This allows for immediate action or decision-making to be made based on the data, without the need for batch processing and waiting for large amounts of data to be collected and analyzed. Streaming data processing is commonly used in applications such as financial transactions, social media, and **Internet of Things (IoT)** devices.

The last type of data pipeline is hybrid data processing, which is a method of data processing that combines both batch and real-time processing capabilities. This allows you to efficiently and effectively handle large amounts of data, as well as quickly access and analyze data in real time. This approach is often used in systems that require both high-speed and high-volume data processing, such as in financial transactions and supply chain management.

The properties of a data pipeline

Based on the functional specifications, as well as the criticality and peculiarity of each data pipeline, there are several important properties to consider when you're designing and implementing a data pipeline:

- **Scalability**: A scalable data pipeline allows you to efficiently and effectively process large amounts of data. It is designed to handle a high volume of data and can be easily expanded to accommodate growing datasets without compromising performance. A scalable data pipeline typically consists of a series of data processing and storage components that are designed to work together to extract, transform, and load data from a variety of sources, such as databases, sensors, and online platforms. It also includes mechanisms for monitoring and managing the data flow, ensuring that data is processed accurately and efficiently.

- **Flexibility**: A data pipeline is often a complex system, but it needs to be flexible in a way that allows for easy modification and adaptation to changing needs and requirements. This property of the data pipeline allows for the integration of new data sources, the incorporation of new analysis tools and algorithms, and the ability to easily change the flow and processing of data. A flexible data pipeline allows organizations to quickly and easily adapt to changing business needs, and to easily integrate new technologies and data sources as they become available.

- **Resilience**: This is a property that is needed when your pipeline has to continue functioning properly, even when one or more of its components fail. This typically involves implementing redundant systems and backup processes to ensure that data is continuously collected, processed, and delivered, even if a failure occurs. The goal is to minimize disruptions and ensure that the pipeline can continue to operate smoothly, even in the face of unexpected issues, without losing data or interrupting operations. This includes measures such as redundant infrastructure, robust error handling, and automated recovery mechanisms to ensure that the pipeline continues to function properly even in the face of unexpected challenges.

- **Idempotence**: This is a very important and often must-have feature of our data pipeline. It implies that the outcome of the pipeline is the same, regardless of the moment and the number of times it is run. This means that if the pipeline is run multiple times on the same input data, the output will be the same each time. Idempotency is a useful property for data pipelines to have as it ensures that the results of the pipeline are consistent and predictable. This can be particularly useful in scenarios where the pipeline is run on a schedule or in a distributed manner as it ensures that the output of the pipeline is not affected by potential race conditions or other issues that can arise in distributed systems.

Rationalizing the costs

At this point, most companies have been building data pipelines for decades, and what initially started as a simple process of transforming and uploading dashboards has now evolved into real data departments with tens, hundreds, and thousands of people working with data. We started by having and maintaining a few pipelines, but today, we have companies with thousands of pipelines that read and write from thousands of different data sources. Therefore, a critical aspect is governing this ecosystem of data pipelines and data stakeholders as well as governing the associated costs. This is especially true when we speak about cloud data architectures based on **Software-as-a-Service (SaaS)** being available on demand, a kind of provisioning well known for being difficult to measure, control, and predict costs.

Due to this, rationalizing data pipeline costs has become not only important but crucial to guaranteeing the right return on investment and making data analysis sustainable for businesses.

Data pipeline costs

Before we understand how to rationalize and keep pipeline costs at bay, we will list and comprehend all the costs that can be generated by correctly and rigorously maintaining a pipeline in a production environment:

- **Infrastructure costs**: This includes the cost of hardware (servers, storage systems, and networking), software, and cloud services needed to build and maintain the data pipeline.

- **Data storage and management costs**: This includes the cost of storing and managing the data that is generated, processed, and stored in the data pipeline.

- **Data processing costs**: The cost of processing and transforming the data, such as using SQL queries, algorithms, and machine learning models to extract insights from the data.

- **Data security and compliance costs**: This includes the cost of implementing security measures to protect the data, such as encryption, real-time continuous vulnerability checks, access controls to protect sensitive data from unauthorized access, data backup and recovery systems, and ensuring compliance with industry regulations and standards with procedures such as data deletion and data retention.

- **Maintenance and support costs**: This includes the cost of maintaining and supporting the data pipeline, including regular updates, technical support, data migrations, and new data integrations.

- **Staffing costs**: This includes the cost of hiring and training personnel to manage and operate the data pipeline (for example, data engineers, data architects, data analysts, data scientists, and IT staff).

- **Site Reliability Engineering (SRE) costs**: This includes the cost of acquiring and maintaining the necessary equipment and software to monitor data and conducting regular audits and assessments to ensure the accuracy and reliability of the data pipelines. What's also important are the costs of storing, analyzing, interpreting, and maintaining the data collected by the data monitoring system.

- **Data analytics and visualization costs**: This includes the costs of the tools and services that are used to analyze and visualize data, such as data mining and business intelligence software.

- **Implementation costs**: This includes the costs of acquiring data from external sources, as well as the costs of integrating this data with existing datasets.

Using data observability to rationalize costs

As we have seen, the costs that are linked to a given pipeline can be numerous and significant. Therefore, any useful support for rationalizing and reducing costs is certainly welcome. Data observability can help in rationalizing costs in several ways. Let's understand how investing in data observability can have a quick return on investment and help rationalize the costs of our data pipelines.

Insights about the performance and efficiency of your data platform

In the previous chapters, we learned that data observability tracks information not only about the data applications but also regarding the context in which the applications run: the system log environment and the software and hardware metrics. Tracking, monitoring, and analyzing this information makes it possible to detect several very useful insights.

One of the most important features that data observability offers is the ability to identify problematic data pipelines with performance degradation or abnormal use of resources. These kinds of anomalies must be detected as soon as possible, especially to avoid an anomaly using the resources that are shared between data pipelines, which can harm the entire data platform. Identify areas where costs may be unnecessarily high or where resources are being wasted, allowing for more informed decision-making and cost-saving strategies.

More and more often, data infrastructures are implemented on the cloud using SaaS, which, among other things, has a pay-as-you-go business model. This implies that using resources incorrectly can have a drastic effect on costs.

Detecting this kind of anomaly can bring about several actions:

- The adoption of data compression techniques to reduce the amount of data being read, processed, and stored

- Utilize data caching and other performance-enhancing technologies to improve data access and retrieval speed

- Use data indexing and partitioning to improve query performance and enable faster access to data

- Review or resize the hardware available to optimize the data platform's performance so that you can identify and address any bottlenecks or inefficiencies

- Invest in training and resources to support the proper usage of the data platform's tools and services

Relying on the power of data lineage

One of the key features provided by data observability is the capacity to infer and build data lineage so that we can always have a detailed overview of how data has been transformed and moved through different systems and processes, including any changes made to its structure or content. Lineage can be tracked at different levels:

- Application lineage
- Data source lineage
- Field lineage

Application lineage is the highest level of data lineage and it is nothing but a representation of your data pipeline, and therefore the graph of dependencies among all the data applications that compose it. One of the main benefits of data observability is having constantly updated and versioned documentation of your data pipeline. You don't have to manually keep track of each change and each new or removed dependency.

This is depicted in the following diagram:

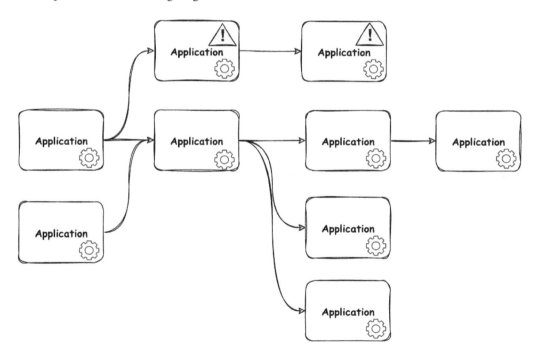

Figure 7.4 – Data lineage – application level

Data source lineage automatically gets full data source lineage, including track back, track forward, and impact analysis, to securely meet compliance/regulatory requirements and easily manage the risk in the pipeline, such as performing change management.

This lineage can be useful for understanding the locations and life cycle of your data during critical tasks such as moving data to a new storage area or while upgrading and replacing data integration tools. Because data lineage provides an overview of how data is produced and propagated through the different areas of the company, it facilitates planning and executing migrations and upgrades. Additionally, it allows data teams to clean up the data system and archive or delete old, irrelevant data sources; this improves overall performance and reduces the amount of data that needs to be maintained, stored, read, processed, and backed.

The following diagram shows a classical form of data lineage that involves different data sources:

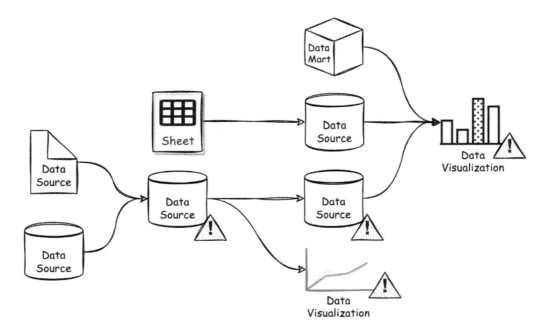

Figure 7.5 – Data lineage – data source level

Field lineage is the lowest and most detailed level of lineage tracking and helps us understand each field of each data source. In particular, it helps us discover and document the dependencies among data sources at the field level. This helps us answer different questions:

- *What are the upstream dependencies of field X?* This question is raised often, especially when there is a suspicious data quality issue in that field and we want to build the dependency tree to speed up the root cause analysis.

- *What are the downstream fields related to field X?* The answer is particularly useful if data quality issues are discovered on field *X* that can potentially be propagated to other fields and data sources. It's also useful to facilitate change management and prevent unforeseen impacts on unknown dependent fields.

- *Is field X completely unused?* No matter how many times we work with legacy and scarcely documented data pipelines and data sources, it's impossible to answer this question without proper lineage at the field level. Not knowing whether a field is used or not makes your data platform substantially immutable and makes it impossible to fight the legacy that's been accumulated over the years.

- *I have sensitive data in those fields. How can I track all the dependent fields that could potentially enrich the sensible data?* Auditing and ensuring that data is stored and processed in compliance with data governance policies and regulations is supported by data lineage as a compliance mechanism. Organizations will find noncompliance issues to be a time-consuming and costly undertaking without data lineage tools.

The following diagram shows data lineage at the field level:

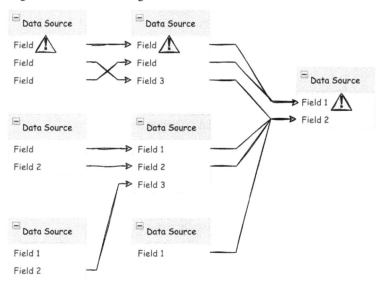

Figure 7.6 – Data lineage – field level

Automating the documentation of existing pipelines

Automatically tracking metadata such as data sources, schemas, and lineages ensures that new data sources are consistently added to the catalog without manual intervention. Using metadata management tools to automatically capture and store metadata for each data source, such as data types, structures, and relationships with other data sources, helps ensure that the catalog is accurately maintained and up to date.

Updating the data catalog can be done manually or through automated processes, and the implementation of automatic data quality checks and alerts to identify and flag any inconsistencies could easily be integrated into the data catalogs. This can help prevent incorrect or outdated information from being used in analyses or decisions. Companies that have this automation in place not only drastically reduce maintenance costs but can rely on data catalogs that are more accurate and less error-prone.

Also, this automation could include the use of data governance tools to manage access to the data catalog and ensure that only authorized users have access to sensitive or restricted data. This helps maintain the integrity and security of the catalog.

All these precautions can drastically reduce the TCO of the data pipelines, especially those that contain the maintenance costs of a data catalog, which increases with the number of projects.

Automatically updating the documentation helps reduce the changing risk. For example, data lineage can provide visibility into the impact of business changes, such as upstream and downstream reporting. Data lineage can be helpful when you're identifying how many tables, dashboards, and reports could be affected by a change in a data element's name and, subsequently, how many and which stakeholders are accessing the report as a result.

Let's summarize what we've learned in this chapter.

Summary

In this chapter, we addressed some very important issues related to data observability. We focused on learning about the main concepts surrounding data pipelines and how they can be characterized, after which we understood the various types of data pipeline architectures.

Then, we learned how data observability can make a drastic contribution to containing and reducing costs associated with the evolution and maintenance of data pipelines.

After, we analyzed and understood the fundamental role of data lineage and when it is essential to automate the documentation updates, reduce data catalog management, anticipate propagation, mitigate the impacts of a data anomaly, and drastically reduce changing risk.

8

Organizing Data Teams and Measuring the Success of Data Observability

This chapter is about how to introduce data observability to your team. It describes the different types of data teams, the different types of organizations these teams need to fit into, and how you can measure the success of this initiative.

First, we will analyze the data team, understand its main roles in detail, and analyze the characteristics of each role and the way they work together. It will also be important to understand how these data teams can be organized and how to better organize the data teams to achieve optimum results. We will see that there are different ways to organize a team, all with their advantages and disadvantages. We will analyze these in detail.

We will also see that these data teams are not easy to manage and that the organization depends on several other factors, such as their size, their maturity, and how the entire company is structured. We will also see that certain types of data team organizations are closely related to the concept of a **data mesh**, which has recently gained interest and traction in companies that invest heavily in data.

In the last part of this chapter, we will learn how to start and spread a data observability initiative in our company and about the important steps to measure the success of this type of initiative.

Accordingly, we will cover the following topics:

- Data teams and their organizations
- How to measure the success of data observability

Defining and understanding data teams

In recent years, investment in data platforms and tools has grown exponentially. At the same time, and proportionally, investment in data teams has increased to the point where the number of data teams is in the hundreds and even thousands.

On the one hand, this has been and continues to be an exciting time for the data ecosystem and for those who work in it, but on the other hand, this exponential growth has also brought with it a whole new set of challenges, not only technical but also organizational. Over the years, several questions have spontaneously risen:

- How can I scale a data team?
- What skills and roles are required for the success of my data investment?
- How does management, such as the hiring process and budget, differ for these specific roles?
- How can I improve communication between my data team and the rest of the organization?

These are non-trivial questions that are difficult to answer. Simply put, we have witnessed the emergence of a new world for which no one was quite prepared.

The roles of a data team

The number of roles is so large that it is often difficult to explain the specifics of each. One of the failure factors of investing in data is to concretely communicate the specifics and potential of each data tool and its role to the rest of the organization. Often, internal stakeholders do not have a clear understanding of their specifics, which increases their mistrust and trust and, most importantly, does not facilitate the optimal use of this great available potential.

So, now is a good time to pause and list the most popular roles in a data team:

- Data engineer and ETL engineer
- Data analyst
- Data scientist and machine learning engineer
- Data steward

Now, let's understand the main capabilities and responsibilities of these roles.

Data engineer

One of the most important pillars of a data team is the data engineers. They often work with large amounts of data and must have a good understanding of various data processing technologies and tools, such as Spark, Kafka, SQL, and NoSQL databases. Data engineers must also have a good understanding of software engineering principles, networks, and system architecture.

The data engineer is responsible for several activities:

- Designing, building, and maintaining the data infrastructure required for data management, data cleaning, and data processing within the organization
- Building, planning, and monitoring the data pipelines
- Implementing complex and sophisticated data storage systems
- Implementing data security protocols
- Automating data transmission and acquisition
- Managing data access
- Assisting other team members in performing complex data collection tasks

Considering the scope and heterogeneity of these activities, a good data engineer needs several hard skills, in addition to soft skills:

- The problem-solving skills required to tackle complex challenges using innovative tools and solutions.
- The continuous evolution of the available solutions and instruments requires a high degree of readiness for continuous learning.
- The data engineer does not work alone; instead, they must deal with different stakeholders and multidisciplinary teams, so good communication skills are also important.
- Understanding the business. Although this is often ignored, it is a critical factor as the data reflects the business of the company. For example, it is impossible to model the data, predict its evolution, or ensure its maximum quality without a basic understanding of the world that generates or uses it.

Sometimes, the role of the data engineer is confused or even interchanged with the similar role of the ETL engineer, but overall, even though some responsibilities overlap, these roles have an important difference: the data engineer is much more focused on creating and maintaining the data infrastructure, while the ETL engineer is more focused on implementing ETL applications that serve to move and transform data. It is also true that, in most cases, the role of the ETL engineer has evolved into the role of the data engineer. Sometimes, the traditional ETL role was simply renamed when the "hype" around data engineering exploded so that the ETL engineer role had more responsibility for maintaining the infrastructure.

The data architect is another role that is often confused or mixed up with the data engineer role. However, in reality, they are two distinct roles: data engineers focus on the technical implementation of data systems and pipelines, while data architects are more concerned with designing the overall data architecture of an organization. This includes defining data models, structures, and relationships for managing data across the enterprise, as well as selecting appropriate technologies and tools for the data platform. Rather than focusing on individual pipelines or the performance of individual tables, data architects take a more strategic approach to managing the entire data infrastructure, ensuring that it is scalable, resilient, and meets the needs of the business.

For data engineers, data monitoring offers fundamental support to monitor and ensure that the data infrastructure and pipelines are working as expected, as well as valuable support to accelerate the process of root cause analysis and problem detection and resolution.

Data analyst

To make your life easier, you can think of the data engineer as the one who provides data, infrastructure, and power, while the data analyst is one of the main roles that leverages all of these functions. You might be wondering which came first. The data engineer or the data analyst?

The role of the data analyst is to analyze data to provide useful information to the internal or external stakeholders of the organization they work for. This includes collecting, cleaning, processing, and interpreting data to create reports, dashboards, models, and forecasts.

Again, this is a role that requires T-shaped skills – that is, individuals who have advanced vertical expertise in one area but are also able to work across disciplines along with experts in other business areas.

Let's look at the main tasks of a data analyst:

- Identify, collect, explore, and analyze data with statistical analysis
- Synthesize data through reports and dashboards to understand the most relevant information
- Design and analyze experiments
- Monitor the quality of collected data to validate the analysis

Data observability helps data analysts constantly monitor the quality of data in real time so that they can automatically detect data incidents. It also offers fundamental support for root cause analysis and resolving data problems.

Data scientist and machine learning engineer

In recent years, the data scientist role has become incredibly popular. So much so that it's even been called "the sexiest job of the 21st century." It's time to understand why, in a way, this is true.

If data teams were initially composed of data analysts and data engineers, at a certain point, the figure of the data scientist became equally important and popular. This happened thanks to the proliferation of machine learning and artificial intelligence tools and models, which made it affordable and convenient to invest time and money in these skills and technologies. With the exponential increase in available data, the importance of data scientists as a strategic role in organizations has also increased as they're responsible for, and enablers in, identifying opportunities to grow the business and improve the product through advanced data analytics.

Let's look at the main tasks of a data scientist:

- First and foremost, an excellent data scientist must have deep domain knowledge because mathematical models are the mathematical representation of the business. Therefore, it's important for data scientists to constantly be aware of the scope, history, vision, and evolving context of the data they are analyzing. So, if you have a data scientist who is also one of your company's first employees, you are in luck.

- They must collect the data required for the analysis – for example, they will query databases with a SQL script or retrieve data from the available APIs.

- They must clean up the collected data to manage missing data and outliers and to standardize the format, which can often be particularly heterogeneous when data comes from different sources.

- They must analyze the available data and identify the statistical technique or machine learning algorithm according to the statistical model's life cycle, a structured process that includes steps such as model development, testing and validation, and deployment.

- They must monitor and review the performance of the model to determine when the model needs to be updated to ensure it remains accurate and relevant.

- Upon completion of the analysis or creation of the model, the data scientist needs to share their results, including limitations and strengths.

The data scientist role can be confused with another role – machine learning engineer. The latter, while part of the data science team, is more focused on building and deploying machine learning models at scale and maintaining the infrastructure required to deploy machine learning models, including data pipelines, optimizing performance, and scaling models. Typically, the machine learning engineer follows DevOps practices that combine software **development** (**Dev**), IT **operations** (**Ops**), and data scientist needs to automate the process of deploying data models and maintaining the data model's infrastructure.

Data steward

The data steward plays a crucial role in empowering the organization to gain a comprehensive understanding of, and derive maximum value from, its data sources. Serving as the data expert, the data steward tries to extract value from data by identifying its users and their motivations for using it and exploring the potential benefits of its utilization. By fostering close collaboration, the data steward is uniquely positioned to propel business success through various actions, including the development of a business glossary, monitoring use cases and data utilization, establishing data quality standards, providing training, and actively participating in governance activities. This role is dedicated to supporting data users, the products that rely on this data, and the strategic application of data, with the overarching responsibility of delivering a robust data management program for a specific data domain while proactively enhancing the data's health within that domain.

Their main tasks are as follows:

- To be responsible for comprehending the data within their domain, including its source, structure, and meaning. They should have a deep understanding of the data to ensure its accurate and effective use.

- They must provide training and support to users and stakeholders to enhance their understanding of data, data tools, and best practices for data usage.

- They must address data-related issues and discrepancies, including investigating data anomalies and working to resolve them promptly.

- They must ensure that data handling and usage align with regulatory requirements and industry standards, particularly in industries with strict data compliance regulations (for example, healthcare and finance).

- They must oversee the entire data life cycle, from data creation and collection to archiving and deletion, to optimize data management practices.

- They must collaborate with IT and security teams to ensure data security measures are in place, including encryption, access controls, and data masking, where necessary.

- They must own and maintain field-level documentation for data sources.

- They must document what the data means, who uses it, how much it is used, and for what use cases.

- They must share data updates and changes with the data consumers.

- They must analyze and resolve requests and ensure timely follow-up and completion as per business rules and **service-level agreements (SLAs)**.

- They must maintain a master list of data quality issues and pending projects/resolutions.

- They must work with key data users to develop business logic for the producers to implement into their systems for automated ongoing monitoring. This helps them proactively identify data quality issues before they impact downstream processes.

Now that we have seen the main capabilities and responsibilities of these roles, it's time to understand how to organize the data team.

Organizing a data team

It is very complex to organize a data team so that it is innovative, cohesive, and resilient while being close to the needs of the business. The recent emergence of these data teams resulted in them often encountering inexperienced management who were unwilling to lead or collaborate with a data team. This is because the roles we learned about in the previous section are often not fully understood, or there is so much confusion, ignorance, and skepticism that misunderstandings can occur between departments and with internal and external stakeholders.

Before you decide how to organize a data team, there are several factors to consider:

- The size of the data team and the available budget

- The size of the company

- How the rest of the company is organized

- The maturity of the data tools and infrastructure

- The maturity of the knowledge and use of the data in the rest of the company

In addition, there are some indicators that you should always keep in mind and monitor to measure the success of your data organization:

- **The satisfaction of internal and/or external stakeholders**: As trivial as it may seem, it is not always obvious that a team is organized to excel in satisfying its internal or external customers since a highly skilled and especially vertical data team may relapse into self-centeredness, making it exceptional but not very useful to the business. However, the success of a team depends on the value it unlocks within the organization, and this should never be forgotten.

- **The quality of the data, analysis, and reports produced by your team**: It is one thing to have one or many satisfied customers, but it is another thing to ensure a very high level of quality and control. Data teams are no longer seen as experimental research and development teams and they must ensure a high level of reliability to avoid making the mistake of making large investments based on incorrect assumptions based on incorrect data or analysis.

- **The efficiency and efficacy of your team in terms of delivery**: It is clear that quality is not enough as a success parameter for a data team, but that it is also necessary to ensure that the team is efficient and effective and that the organization enables the team to express its potential in the best possible way to ensure a flexible and efficient service within the company.

- **Innovation**: In addition to the various parameters that must be considered for the team to be fast, reliable, and able to meet a variety of needs, it is also important for the team to have skills, tools, and an infrastructure that is constantly updated with the technologies that the market and the open source world are producing. The data ecosystem is still in a phase of evolution and adaptation, new solutions are on the agenda, and a team must be able to explore and use the most effective solutions. This is a very important factor to consider because, as we will see, some organizations reward or penalize technology engagement.

- **Team bonding, team satisfaction, and people retention**: A team also needs cohesion, vision, and support. As we will see, some organizations reward efficiency but do not value team building or encourage the growth of individual team members. Professionals who work with data are increasingly in demand, so retaining this talent is not negligible for an organization that truly views data as a business asset.

There are three main types of data team structures:

- Centralized or isolated

- Decentralized or integrated

- Hybrid

Let's analyze the pros and cons of each organization based on the aspects we have analyzed so far.

The centralized data team

The first organization we will analyze is the so-called centralized team, also known as the isolated team approach. This type of team is essentially a single data team that contains all of the data team roles we listed and analyzed earlier in this chapter:

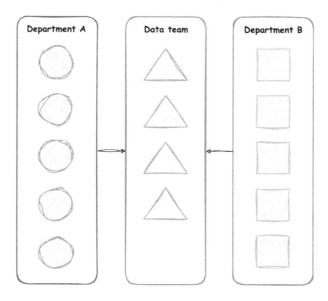

Figure 8.1 – The silos of a typical centralized approach

The centralized data team has the following characteristics:

- They own the data completely in terms of access, quality, life cycle, platform, analysis, and more.

- They have their own data roadmap and focus on implementing the key points of that roadmap. For example, they consider the enhancements and developments needed for the data platform, or the ongoing research development needed to explore and test new tools and new technologies.

- This centralized approach often requires a dedicated data project manager to focus on that team's roadmap and vision.

The centralized approach offers the following advantages:

- It's easier to focus on the data roadmap, so it's easier to promote and plan new data initiatives such as launching a new data warehouse or refactoring some legacy data pipelines.

- Since all members of the data team work closely together, there is usually more individual development, especially for the younger profiles who have the opportunity to work with other data analysts, data scientists, or data engineers with more experience.

- When everyone works together, data people speak the same language and don't have to spend time and energy trying to convince others that certain initiatives make sense. This fosters team building and also more engagement and commitment.

- It should prevent duplication of work across different teams

There are also important disadvantages to consider with this approach:

- Because you have only one central data team across the organization, you must coordinate all requests from external stakeholders and departments. This is particularly challenging because your resources, or at least some of them, are centralized but also used by other teams that may be highly dependent on them. The data team can easily become a bottleneck in this case, and it is extremely complex to coordinate your data roadmap with the roadmaps of all the other teams. For example, remember that all teams need to set **objectives and key results (OKRs)** for the next quarter, and every time this happens, you need to do the following:

  - Set the OKRs for your data team to implement your data technical roadmap

  - Listen to all the OKRs from the other teams and identify and prioritize all possible connections and dependencies

  - Identify and prioritize the activities of your team and stakeholders

- Another problem is that this organization does not help spread the data culture and its potential throughout the company – that is, a rather isolated team does not help spread and share all the knowledge and potential of the team and the data skills of the team. So, there is a risk that this kind of isolation from other cross-functional teams, and especially far away from the business, prevents a cultural exchange that can trigger very important projects and ideas.

Here are some examples of when it's a good idea to take the centralized approach:

- You have limited resources and a small data team of 6–7 people at most, and if it is understaffed compared to the needs of your business.

- If the rest of your organization is organized in silos – for example, a marketing team, a frontend team, a mobile team. In this case, it makes less sense to decentralize your team.

- Your data team is still in the start-up phase – for example, if you do not have a solid data platform yet or if your data department has not implemented defined processes and best practices yet.

- Your data team is young and consists mostly of junior members.

The embedded data team

A second common method for organizing a data team is the integrated or embedded approach. In this type of approach, there is no longer a central data team; instead, the data team members are distributed throughout the organization and embedded in the other departments or the other cross-functional tribes and groups, meaning that in this case, the responsibility for the data is not centralized in a single data team and is embedded in each group. In these cross-functional teams, there are no longer just data team members, but all the roles required for the cross-functional team.

Let's say we're in a cross-functional product development team. It must have all the roles it needs to achieve its goals and be as independent as possible from the other teams, and it owns the priorities of the data team members, so it is responsible for prioritizing the team's tasks, including data:

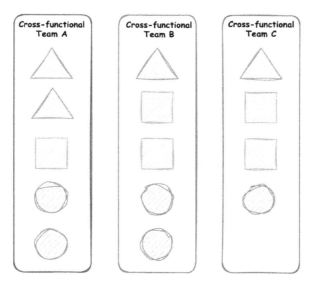

Figure 8.2 – The typical embedded approach

Let's talk about the advantages of this approach:

- First, collaboration with the other business units is increasingly encouraged as responsibility for data and data quality no longer rests with a centralized data team but with a cross-functional team that is not only the data producer but also the owner and is responsible for the quality of that data.

- With this organization, the team might be able to identify and resolve data issues more easily because, as part of the embedded team, they are more aware of the impact on the business and, as data producers and data owners, can focus more on resolving these types of data quality issues as quickly as possible.

- Another important benefit is that this approach also promotes the democratization of data because data ownership is embedded in the team, which makes it easier for the team to access, fully know, and understand the data being produced and consumed. So, ultimately, this approach promotes the spread of data culture.

- It enables data to become truly central to the business. Data is no longer an asset that resides in one part of the organization but is now understood – it is now distributed throughout the organization

- Last but not least, this approach is extremely scalable because, as your data team grows, you need to distribute your data members across the various cross-functional teams that the organization needs to form to scale the entire organization.

Despite the many advantages, managing an embedded business unit also brings with it some disadvantages and new complexities that need to be addressed:

- One of the major drawbacks is that it is more complicated to ensure the progress of the long-term data roadmap. For example, because all data engineers are embedded in different cross-functional teams, there is a risk that they will be more focused on helping the team with their immediate needs, so they are more likely to focus on short-term and mid-term issues rather than the long-term data roadmap.

- Another problem is that this type of approach can be less efficient because, for example, there is a need for better coordination and communication between data team members spread across the organization, which can sometimes lead to several other data engineers doing similar things in parallel. There is also a risk that someone will reinvent the wheel because of a lack of documentation or communication between data team members spread across different cross-functional teams.

- In this approach, it's critical to define clear and common data governance processes that must be implemented and followed throughout the enterprise. So, from this perspective, the company needs to invest in budgets, people, and time because as important as it is that data ownership is distributed throughout the company, it's also true that data governance must be followed properly so that you avoid losing control of what is happening.

- Working in teams that aren't as focused on data can be frustrating because they can't keep up with colleagues and managers from other disciplines, or because they can't focus on long-term projects, which are often also the most exciting challenges. That's why management needs to respect the priorities and importance of data when data professionals work in cross-functional teams. In addition, it can be more difficult for junior profiles when data colleagues are spread across the organization rather than having them on a single team.

Now, let's look at some examples of when it's a good idea to adopt the embedded approach:

- The data department has already laid its foundations, so you already have consolidated processes and a mature and reliable data platform and tools in place.

- Your data team needs to focus more on the needs of the business and product development teams, and you decide to focus less on the long-term data roadmap to support the cross-functional teams.

- Your company is already organized with cross-functional teams – for example, your product development teams are already organized with tribes and squads.

- Your data team consists of an interesting number of members – for example, more than 20. In this case, you need to implement a scalable organization, and this is certainly one of the best approaches you can take.

- The team mainly consists of senior members, so the average seniority is high. This is important because, as we analyzed earlier, working in cross-functional teams is more challenging for junior profiles. In any case, the younger members of the team will have to adapt to this approach if it is adopted. Older employees will leave the company and take their expertise with them – this is inevitable. So, to maintain a talent pipeline, you need to recruit at a younger level.

The hybrid data team

We have already analyzed that the centralized and embedded approaches have different limitations and that they are still far from being perfect. For example, one of the main limitations of the embedded approach is that it might be very difficult to maintain focus on the data roadmap.

For this reason, there is a third common approach called the hybrid approach. The difference between this approach and the embedded one is that you still have embedded data experts in the cross-functional teams, but you also have a dedicated central data team that acts as a buffer for the cross-functional teams when they need additional help and also helps those cross-functional teams implement complex features and complex tools. So, the idea is to relieve the cross-functional teams from building and maintaining complex data projects and allow them to focus on what is more pressing for the team. Of course, the central data team will also focus on the data roadmap and work on more strategic projects that will help the entire business:

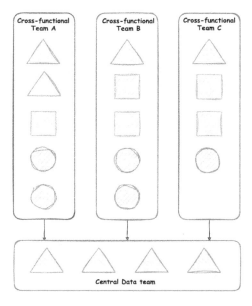

Figure 8.3 – The hybrid approach

The main characteristic of the hybrid approach is that it inherits all the most important pros and cons from the embedded organization – in the end, it's just its extension. Also, with this approach, the owner of the data remains the cross-functional team, as is the case with the embedded approach, but there is a big difference in that another data team exists that's fully dedicated to the data technical roadmap and more focused on establishing and executing the technical roadmap. For example, they own the support teams, so they could also work on challenging data projects that cannot be handled by the cross-functional teams because they all have their priorities.

There are many advantages to this approach, and most of them are in common with the embedded approach. However, a dedicated central data team can have resources that focus on implementing your data roadmap while other ones are still embedded in the cross-functional team.

Of course, in this case, the big concern is that we need a bigger investment because we have to build a dedicated team. We can say that, in some cases, it could be the best solution but at the same time the most expensive one.

Data mesh, data quality, and data observability – a virtuous circle

Data mesh, data quality, and data observability are three very important components that can help you build a robust and effective data strategy in your organization since each component plays a specific role in ensuring that data is accurate, consistent, and available. All of these components can ensure that your organization can make informed decisions and base your decisions on data that is not only available but also as accurate as possible.

By working together, these three components – data linkage, data quality, and data observability – can create a virtuous cycle that builds confidence in the data and the strength of your data infrastructure and architecture, leading to better outcomes for your data teams and, of course, all stakeholders who rely on your data teams and outcomes.

To understand how and why these components can and must work together, it's important to know what data mesh is and look at its core principles.

Data mesh

Data mesh is a paradigm for building data organizations and data architectures that emphasizes the importance of decentralizing domain ownership. This is related to the decentralized organization of data teams that we saw earlier in this chapter. There, we saw that in this type of organization, data is not owned by a central data team; instead, it's owned by individual business units or individual product groups or domains. So, in this case, the domain is responsible for data production and ensuring the quality, availability, and usability of its data.

Data mesh includes several principles:

- **Data as a product**: Data should be treated as a product that is created and used by other teams in the organization. This means that data should have clearly defined interfaces and be designed to meet the needs of users.

- **Self-organizing teams**: Cross-functional teams should be organized around specific business units and be responsible for the end-to-end delivery of data products. This promotes accountability and ownership of data in each area.

- **Domain-based design**: Data architecture should be designed by the business domain rather than the technical domain. This promotes alignment between the data architecture and the business needs of the organization.

- **Federated data management**: Data management should be decentralized and in the hands of the domain teams. This promotes flexibility and agility while ensuring that data is managed responsibly.

- **API-centric design**: Data products should be developed with APIs that enable easy integration with other systems and applications. This promotes interoperability and reuse of data products.

- **Self-service data infrastructure**: Business teams should have access to a self-service infrastructure that's managed by a centralized data team that allows them to manage their data pipelines and data. The data infrastructure should be designed as a platform that provides reusable services to domain teams, promotes agility and standardization, and reduces duplication of effort across the enterprise. Standardization promotes observability by providing a consistent way to monitor and track data across the organization and enabling teams to easily monitor their data products.

- **Data product life cycle**: Data products should go through a life cycle that includes ideation, incubation, and production phases. This fosters a culture of experimentation and innovation while ensuring that data products are well-designed and meet user needs.

Building the virtuous circle

Data quality, on the other hand, refers to several properties, such as the accuracy, completeness, and consistency of your data. Thus, good data quality ensures that the data is reliable and can be used for decision-making. In a data mesh environment, data quality should be owned and managed by the subject matter teams. This promotes accountability and ensures that data products are reliable and trustworthy, which means there is a built-in incentive to ensure high data quality. This focus on data quality leads to better results for the area, which, in turn, benefits the business as a whole.

On the other hand, data observability is the ability to monitor, instrument, and understand your data, and it enables your organization to proactively identify data issues and understand how to fix the root cause of the problem and the data itself as quickly as possible. Data observability allows organizations to monitor how data flows through their systems and identify issues before they become problems. This allows organizations to quickly fix problems, which helps them maintain data quality and ensures that data is available when it's needed. By monitoring data, companies can identify trends and patterns that can be leveraged to improve business outcomes.

Taken together, data mesh, data quality, and data observability form a virtuous cycle. By encouraging domain ownership, data mesh creates incentives for domains to focus on data quality, leading to better outcomes:

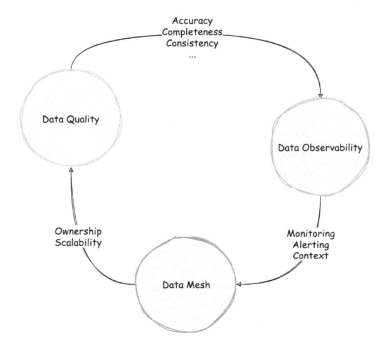

Figure 8.4 – The virtuous circle of data mesh, data quality, and data observability

Overall, the virtuous cycle of data mesh, data quality, and data monitoring helps build a strong data infrastructure that can meet the needs of the business. The interaction of these three components fosters a culture of data-driven decision-making that enables organizations to make better decisions, improve outcomes, and drive growth.

Data observability can play an important role in your data mesh initiative, especially when it comes to domains within the data mesh infrastructure. You may have wondered who in your organization should be responsible for the data observability process. Usually, you will achieve the highest effectiveness if you assign this responsibility to the domain owners. Since they're responsible for their data, the producers will be more likely to monitor and observe their data closely, and can quickly identify and address any issues that arise.

Now, let's learn how to trigger a data observability initiative in a company.

The first steps toward data observability and how to measure success

Implementing a data observability initiative in an organization requires performing a series of steps. This section will provide general guidelines to help you get started.

First, you have to identify and involve the stakeholders who will be affected by the data observability initiative. This may include data analysts, data engineers, business analysts, and other members of the data team.

After detecting the stakeholders, it's time to define the clear scope of this data observability initiative. This may include what data sources to include, what types of data quality issues you want to address, and what metrics you will use to measure success. Usually, the best way to do this is to define the KPIs for measuring the success of the data observability initiative. You must define the KPIs that will be used to track the effectiveness of the initiative, such as the number of incidents resolved, time to resolution, and impact on data quality.

Once you have identified the stakeholders and defined the scope, you need to define the processes for data observability. This may include processes for data monitoring, anomaly detection, data quality assessment, and incident management. It is also important to assemble a team or identify the right roles and responsible parties for such an initiative, such as data engineers, data analysts, and other data team members, who have experience with data monitoring and incident management.

It is also important to develop or select the right tools. There are several commercial and open source tools for data observability, and you need to select the ones that best meet your needs:

- With real-time monitoring of data flows, data processing, and data transformations, the tool should also be able to detect anomalies and trigger alerts when data quality issues are detected. It should be able to profile the data to identify issues such as missing values, duplicate records, or data inconsistencies. It should also be able to identify relationships between data elements and analyze data dependencies.

- End-to-end data lineage allows you to track the origin and flow of data across different systems and processes. We already know this feature is critical for root cause analysis and incident management.

- It should be able to integrate with other data management systems such as data catalogs, data integration tools, and data warehouses so that you can easily ingest data and share the outcome of data profiling.

- It should provide intuitive visualizations and reports that allow users to easily understand and analyze data quality issues and discover the root causes.

- The tool should adhere to security and compliance standards such as the **General Data Protection Regulation (GDPR)**, the **Health Insurance Portability and Accountability Act (HIPAA)**, and **System and Organization Controls 2 (SOC 2)**.

Overall, launching a data observability initiative requires careful planning, coordination, and execution. By following these steps, you can ensure that the initiative is successful and adds value to your business.

Measuring success

Having established the plan, we now need to measure the success of our data observability initiative. While this can be difficult, it is important to understand the effectiveness of the initiative and identify areas for improvement. Here are some ways to measure the success of a data observability initiative:

- **Reduce data incidents**: One of the main goals of a data observability initiative is to reduce the number and severity of data incidents. You can measure the success of the initiative by tracking the number of incidents before and after the initiative is implemented, as well as the severity of those incidents. Ideally, you should see a decrease in both areas.

- **Time to resolution**: Another important metric you should track is the time it takes to resolve data incidents. The faster you can identify and resolve issues, the better. You can measure the success of the initiative by tracking the average time it takes to resolve incidents before and after the initiative is implemented.

- **Data quality**: Data monitoring initiatives should also improve data quality. You can measure the success of the initiative by tracking metrics such as the completeness, accuracy, consistency, and timeliness of data. You can also measure the impact of the initiative on downstream systems that rely on high-quality data.

- **Adoption**: A successful data observability initiative requires buy-in from stakeholders across the organization. You can measure the success of the initiative by tracking adoption rates among data analysts, data engineers, and other stakeholders. You can also track feedback from these stakeholders to identify areas for improvement.

- **Return on investment** (ROI): Finally, you can measure the success of a data observability initiative by tracking its ROI. This can include factors such as the cost of implementing the initiative, the cost savings from reducing data incidents, and the value of improved data quality.

Overall, measuring the success of a data observability initiative requires a combination of quantitative and qualitative metrics. By tracking these metrics over time, you can evaluate the effectiveness of the initiative and make data-driven decisions for improvement.

Summary

In this chapter, we took a deep dive into how data teams have evolved by understanding the various roles and their responsibilities.

We also faced and understood how complex it is to organize these teams, which often work horizontally within the entire company.

We also learned about the key factors for the success of these teams as well as the key factors for the success of large and complex initiatives, such as the introduction of a data observability project in the company.

Part 4: Appendix

The concluding part of the book provides practical guidance for implementing data observability. It presents a comprehensive checklist method to integrate data observability into company pipelines, highlighting common pitfalls and concerns encountered in various implementations.

The book wraps up with a roadmap for data engineers, offering a technical pathway to effectively implement data observability in an initial project and then scale it across the organization. This final chapter serves as a strategic guide to embedding data observability into the core of data engineering practices.

This part has the following chapters:

- *Chapter 9, Data Observability Checklist*
- *Chapter 10, Pathway to Data Observability*

Data Observability Checklist

We have seen that data observability has the capability to become an important component in your data projects. It enables organizations to gain real-time insights into the behavior of their systems and infrastructure, allowing for faster troubleshooting, improved performance, and better decision-making.

However, implementing data observability can also present a number of challenges and pitfalls. After having conducted several data observability projects across different industry sectors, we wanted to share with you some points of attention to make your journey toward data observability successful.

In this chapter, we will explore the perks and drawbacks of data observability, providing guidance on how to implement it effectively while avoiding common pitfalls. We will see how data observability projects may fail, and what the strategies to overcome this are.

By the end of this chapter, you will have a deeper understanding of the challenges of data observability implementation, and we will provide a checklist describing the steps needed to implement it successfully. This chapter covers the following topics:

- Challenges of implementing data observability
- Checklist to implement data observability

We hope that the insights and guidance provided in this chapter will serve as a valuable resource for those seeking to harness the transformative power of data observability, fostering a culture of continuous improvement and innovation within your organizations.

Challenges of implementing data observability

In this section, we will describe the common pitfalls and challenges of the implementation of data observability and how we can overcome them. The concerns we will cover are the following:

- Costs
- Overhead

- Security
- Complexity increase
- Legacy system
- Information overload

Let's start with the bottom line: the costs.

Costs

Foremost among the concerns surrounding data observability are its associated costs, which can pose a significant financial burden on data projects. These expenses typically encompass the following:

- The acquisition or development costs of a data observability solution, including the investment in research and development and the requisite team training
- Expenses related to the storage and computation of data observations, which can also introduce overhead, as we will elaborate on later in this chapter
- The marginal cost incurred when integrating observability into supplementary data pipelines, necessitating operational efforts for implementation and maintenance

To overcome the cost challenges, organizations can plan to use an open source tool, reducing the costs of development and maintenance. Additionally, organizations can explore flexible pricing models from vendors that might offer pay-as-you-go or tiered pricing, aligning costs with actual usage.

Moreover, exploring cost-sharing initiatives across projects or departments within your organization can also help optimize resource utilization. By collaborating on data observability solutions, you can distribute costs more effectively and foster a culture of data-driven decision-making that benefits the entire organization.

Depending on your organization, you may also want to host the solution on-premises or in the cloud. Some solutions to reduce the load of the resources needed will be discussed in the *Overhead* section.

Last but not least, the costs of data issues far outweigh the cost of implementing data observability. The data team can demonstrate the return on investment in order to justify the costs.

To sum up, data observability can come with significant costs, such as the cost of acquiring or developing a solution, storage, and computation of data observations, and the marginal cost of adding it to a supplementary data pipeline. To mitigate these costs, organizations can consider using open source tools and hosting the solution on-premises or in the cloud. Additionally, techniques can be implemented to reduce the load on resources.

Overhead

Implementing data observability may introduce a certain level of overhead into the business logic of the applications it monitors. Indeed, implementing data observability at the source may increase the load of the job or application in the system, taking up additional time and resources in the organization.

As a reminder, data observability is more powerful when data observations are collected as close to the application runtime as possible. If the data observability tool is included in the job, script, or application, it will use its internal resources to compute metrics and KPIs. If data observability acts as an external component, dedicated resources need to be configured.

In addition, the volume of data plays an important role in the overhead. In a data pipeline, if the time consumed by a data observability agent is 15 minutes per application, you may end up with a final output delayed by several hours. Regarding the type of pipeline involved, it may lead to friction with the rest of the team and stakeholders and impact the overall business activities.

To minimize this impact, we recommend several techniques that can be included in the application or the data observability agent.

First, the pipeline can be tuned to start the following step once the business logic of the previous one is accomplished. This will allow the business logic to be completed while the observability tasks are completed in the background. To avoid a dirty read, this tactic will work efficiently if the data source is not modified by successive applications. Resources can also be optimized, allocating less CPU or RAM once the business job is executed, limiting the impact on concurrent jobs.

Second, we can act on the volume of data being processed. When computing the observations, we can filter on a subset of the dataset, but also aggregate or approximate some values. It is quite relevant to do so in applications processing the same data every day.

In summary, implementing data observability can introduce overhead with the business logic of the applications it monitors. This can be due to the added load on the system as well as the need for dedicated resources. To minimize this impact, techniques such as optimizing pipeline timing, allocating resources, and filtering or aggregating data can be used. Additionally, it is important to consider the volume of data being processed and its potential impact on the overall business activities.

Security

Concerns about the new topic of data observability are often expressed by the security team. Indeed, observability requires that the tool has access to a large volume of data, which may of course have a sensitive or private character.

Observability tools, because they summarize a lot of data pipeline information, become a target of choice for hackers. Therefore, it is of high importance to control the data sent to the tool, its nature, and who can access it.

First of all, the data sent for observability must be composed of metadata – "data about the data." In no case should the metadata sent include sensitive or private information. To avoid this, you may use encryption or aggregation. The metrics you compute can be valuable for monitoring or troubleshooting but need to disclose minimum information about the data. For instance, favor a metric computing the number of distinct categories instead of listing those categories.

Second, the observability tool you use must reflect the same segregation as the data source's filesystem. You shouldn't be able to access more data than what you can actually query in the enterprise ecosystem. Access control is an important security component of the data observability system.

Third, any tool you use should be installed in your data system. If a tool needs access to your data, a proxy should be included to avoid sending sensitive information outside of the enterprise system. This is described in *Figure 9.1*:

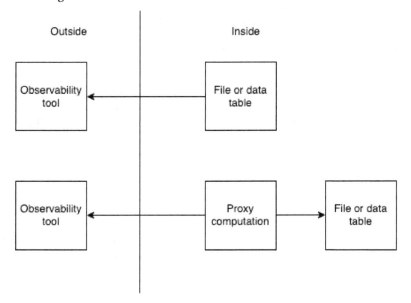

Figure 9.1 – Observability tool and data access

Figure 9.1 describes how we can avoid sharing sensitive information when the observability tool is cloud-based. To avoid a third-party and external application having access to the data, proxy computation can be used in order to access the data. This element, installed on the enterprise system, avoids sharing the data directly with the observability tool. Instead, the proxy computation accesses the data and only shares the results, the computed metrics, with the tool.

In summary, data observability can raise security concerns for organizations, as it involves access to large volumes of potentially sensitive data. To mitigate these concerns, it is important to control the data sent to observability tools and ensure that it does not include sensitive or private information. This can be achieved by using encryption or aggregation of data, and by computing metrics that disclose minimal information about the data. Additionally, access controls should be implemented to ensure that users can only access the data they are authorized to view, and that any observability tools used are installed within the enterprise system to avoid sending sensitive information outside of the organization.

Complexity increase

Data observability may add weight to the architecture of a data pipeline. Complexity is a challenge often met in data observability, as it requires a new tool to collect, process, and analyze large amounts of data from multiple sources. It may lead to challenges in configuring and maintaining the observability tools.

Complexity can arise from the need for multiple tools or techniques to implement data observability, each one with its own specific use case, configuration, and learning curve. This can make it hard to ensure that all of the tools are working together seamlessly and can lead to challenges in troubleshooting and maintenance.

In order to minimize these risks, it is important to consider data observability as an organization-level project, with a dedicated (support) team responsible for its good implementation. If this is not possible in your organization, you could think about training a team member in order to create a data observability champion.

Legacy system

Implementing data observability can be a less arduous task when executed within a new data pipeline as opposed to a legacy system. The challenges posed by legacy systems are multifold.

In a legacy system, the existing architecture and accompanying infrastructure may not be inherently designed to handle the additional load and resources necessitated by data observability. This mismatch can necessitate substantial modifications to the existing system to ensure seamless integration. Furthermore, legacy systems often lack the necessary data governance and security controls required to handle sensitive data in a manner consistent with best practices, thereby making the secure implementation of data observability an uphill battle.

On the other hand, introducing data observability into a new pipeline can be more straightforward. These pipelines can be designed and constructed with data observability in mind from the outset. Consequently, they can be optimized to gracefully accommodate the increased load and complexity that data observability entails. Furthermore, the necessary data governance and security controls can be integrated into the pipeline's foundations right from the beginning. New pipelines also benefit from the utilization of more recent technologies and solutions, further simplifying the implementation of a data observability tool.

From our experience, we strongly recommend that data observability be incorporated as a fundamental step in the creation of new pipelines. This integration process should become a matter of course within the organization. The presence of a data observability agent should be viewed as an essential validation step for each new project. This strategic inclusion ensures that any data issues can be swiftly pinpointed and resolved, fostering an environment of trust in your data.

Furthermore, the same principle should be extended to legacy projects undergoing updates, such as migrations to new libraries or technologies. These updates present opportune moments to introduce data observability into older pipelines and enhance their reliability.

Information overload

With the proliferation of metrics and metadata collected through data observability, organizations often grapple with information overload. This inundation can render troubleshooting and issue detection a daunting task, buried beneath a mountain of variables.

Alert fatigue is one common repercussion of this information deluge. An overabundance of detected anomalies can lead to alert fatigue, where the data team begins to disregard alerts from the system due to their sheer volume. An efficient recommendation system must strike a balance, as an excess of detected anomalies increases the risk of false positives, causing the data team to overlook genuine issues.

Moreover, the excess of metrics can compound the complexity of root cause or impact analysis. When investigating an issue, a surplus of metrics can make it challenging to identify the root cause, obscuring the signal amid the noise. The time and effort required to analyze this extensive data can be substantial and may be challenging to justify.

To mitigate these issues, organizations can employ techniques such as data visualization, facilitating the interpretation of complex data. Additionally, observability tools should harness the context of data observation collection to guide users through the analysis process.

To avoid alert fatigue, the following procedures can be implemented:

- Configure notifications only for alerts that have an impact on other data sources or occur within a scope where action can be taken. Leveraging data observability and its contextual information enables the selective treatment of relevant alerts.

- Regularly review the suggestions from your data observability tool. This will reduce the risk of having false positives over time.

- In the event of a prolonged issue resolution, consider temporarily pausing the rule generating the alert. This prevents the generation of alerts for known issues, ensuring the team's focus remains on unresolved problems.

In this section, we have seen the most common challenges we have encountered while implementing data observability. Here, we listed the costs, overhead, security, complexity, legacy, and information overload issues. Nevertheless, none of them outweigh the benefits of data observability. These challenges should be viewed as points of attention, guiding organizations in selecting the right approach to data observability. In the next section, we will distill these insights into a practical checklist for the implementation of data observability.

Checklist to implement data observability

In this section, we will delve into a comprehensive list of considerations to keep in mind when you embark on the journey of implementing a data observability solution. These questions will not only guide you through your initial project into data observability but also prove invaluable as you progress to more advanced implementations. By carefully addressing these considerations, you will be able to establish a robust foundation for your data observability initiative, one that not only aligns with your organization's objectives but also harnesses its full potential for maximum benefit.

The questions we need to answer are the following:

- Which pipeline should I select to start with the implementation?
- How many applications should I include in the scope?
- What criteria are important to select the observability tool?
- How do we define the set of metrics we want to track?
- How will alerts and notifications be configured?
- How will the data be collected?
- What metrics should I follow to assess the data observability ROI?
- What should I do with new projects?
- How should I deal with legacy projects?

Start with the right data or application

1. Which pipeline should I select to start with the implementation?

The pipeline you first select is a major driver of success for conducting a data observability implementation. To make sure this pipeline can serve as a model of success for further implementations, it is up to you to decide, following your needs, to scope the first project on a legacy pipeline or to initiate observability in a new project.

The advantage of selecting a well-known pipeline in your team is that your assessment of data observability will be made easier as it will be smoother to evaluate the impact on common use cases that you've already experienced within this pipeline – for instance, a pipeline generating a lot of null values while it was not expected and that generated friction with the marketing team.

2.   How many applications should I include in the scope?

When determining how many applications to include in the scope of your data observability project, it's crucial to follow a strategic approach. We recommend a progressive approach that starts from the application responsible for generating the final data source consumed by your organization and works backward through the input steps. This approach offers several advantages:

- **Quality assurance**: Starting with the final application ensures that you assess the quality of your ultimate data output before it reaches stakeholders, and thus the business. This process establishes immediate trust in your data's reliability and integrity.

- **Resource optimization**: Given that time and resources are often limited, beginning with the application closest to the data's ultimate use optimizes the allocation of your efforts. This approach maximizes the impact of data observability on critical data sources.

- **Documentation of application types**: While progressing through your applications, it's essential to keep comprehensive notes regarding the language or type of applications you are integrating. This documentation will prove invaluable as you proceed with the selection of your observability tool.

Once you've defined the scope of applications, the next step is to delve into the selection of the appropriate observability tool.

Choosing the right data observability tool

3.   What criteria are important to select the observability tool?

Now, let's talk about selecting the perfect observability tool for your data projects. Selecting the perfect observability tool for your data projects is a crucial decision, akin to choosing a trusted ally for a challenging quest. We'll break down, in no particular order, the essential criteria you should consider.

Security and compliance

The selected tool must allow the user to comply with the security requirements of the organization. The ideal tool should provide the organization with the same segregation as in their file access.

Ensure that the tool aligns with data compliance and regulations relevant to your industry and region.

Costs

Now, let's talk finances. Cost considerations are always in the mix. You have a few options here:

- Open source versus vendor solutions: You can go the open source route to keep costs down, or you might consider vendor solutions, which can provide a more comprehensive package, but at a price.

- In-house development: Alternatively, you might consider developing an in-house solution. But keep in mind that in-house development might take a bit longer and could be a wild card in terms of cost control.

- Hybrid approach: Another path to explore is a hybrid approach. You could use vendor technologies to collect the essential data while tapping into your internal know-how for data processing. It's all about finding that sweet spot between external tools and cost control. However, we are deeply convinced that the core value of a data observability platform lies in the insights it provides, and not in the integrations it offers.

- Infrastructure matters: Speaking of cost, your infrastructure choice can make a significant difference. Going for a cloud-based setup can be cost-effective, but it might raise some security eyebrows. It's a balancing act, for sure.

Integration

Now, let's get into how smoothly this tool plays with your existing setup:

- Data observability agents: Your chosen tool should have the right agents that match seamlessly with the applications and data sources you're dealing with.

- Interoperability: It's not just about the tool itself but also how it fits into your toolkit. Look for a tool that can integrate smoothly with your data catalog and IT service management tools. This harmony saves you time and keeps things running smoothly.

- Scalability: You might consider whether your toolset could increase in scope in the future, and whether your tool selection would support that.

Data retention

Depending on the pipeline you will monitor, data retention can be an important factor in selecting the right tool. Indeed, especially if your pipeline is running frequently or if it uses streaming data, you may need to keep some historical data in order to start the analysis or to be sure you can address the issue on time.

Intelligence engine

Next, the brains behind the operation – the intelligence engine. This is where the magic happens. Your observability tool should be like a genius data detective, turning raw data into actionable insights. It's not just about having logs, traces, and metrics; it's also about making sense of them. So, choose a tool with a powerful intelligence engine that can unlock the hidden gems in your data.

Customization

Being able to customize the observability tool means that you can adjust its features and settings to suit the specific needs of each project or pipeline. This adaptability allows you to fine-tune the tool's capabilities, ensuring it aligns perfectly with the requirements of the project at hand.

For example, one project might require real-time data monitoring with a focus on specific metrics, while another project might involve batch processing of large datasets. With customization, you can configure the observability tool to excel in each scenario, just as you would select the right tool from your toolbox for each type of job.

Other considerations

Beyond the core criteria we've discussed, there are additional factors to consider when embarking on the journey to find the ideal data observability implementation. These factors can significantly impact your experience and success with the tool:

- User-friendliness: The ease of use and user-friendliness of the observability tool should not be underestimated. A tool that offers a seamless and intuitive interface can save you valuable time and reduce the learning curve for your team.

- Vendor reputation: It's wise to investigate the reputation of the vendor providing the observability tool. A vendor with a solid track record of reliability, innovation, and customer satisfaction is more likely to provide you with a dependable and continuously evolving solution.

- Support and community: The availability of support and an active user community can be invaluable. A responsive support team can assist you in overcoming challenges and addressing issues promptly. Similarly, a vibrant user community can offer insights, best practices, and solutions based on their experiences.

Selecting the metrics to follow

This point is about introducing a checklist to the metrics you need to follow and display to make them observable.

4. How do we define the set of metrics we want to track?

It all starts with a crucial step – establishing clear **service-level agreements (SLAs)** and objectives with your data consumers. This is a collaborative process where you engage in meaningful discussions to understand their needs and expectations.

Once the objectives are well-defined, you can turn your attention to the metrics that will serve as your guiding stars. These metrics are not just numbers; they are the pulse of your data pipeline. You cannot afford to overlook this during the implementation process.

5. How will alerts and notifications be configured?

Following the metrics you want to track, some of them can be used to create rules that will support the objectives. These rules will be used to create alerts on the health of your pipeline.

However, it's important to maintain a balance to prevent alert fatigue. To address this, we recommend a focused approach. For each objective you've defined, consider creating a dedicated rule. The rule must be regularly reviewed by the data or the process owner. The more targeted the rules are and the more accurately they are sent to the right actor, the less friction they will create, making it easier for data observability to be adopted at scale.

6. How will the data be collected?

You have to decide here which trade-off you want to have between the data creation runtime and the observations collection. Synchronicity may require higher resource consumption during the job. However, you can be deprived of valuable information. This trade-off has already been discussed in *Chapter 3*, *Data Observability Techniques*.

The context has to be defined at the start of the application execution in order to frame the observations that will be collected for the application. This context includes the data observability elements described in *Chapter 4*, *Data Observability Elements*.

Compute the return on investment

At the end of your first project, you can already assess the impact of data observability. Though it is a complex process, a meticulous and structured approach can help in accurately gauging the benefits and the overall impact on the organization.

7.   What metrics should I follow to assess the data observability ROI?

Firstly, it's crucial to identify the tangible benefits experienced during the project, such as improvements in overall data quality and reductions in data incidents. Data observability aids in reducing the **mean time to detect (MTTD)** an issue, the **mean time to resolve (MTTR)** the issue, and the time spent on documenting the pipeline, which are pivotal metrics to consider.

To assess the ROI for data observability, start by calculating the return. This involves identifying specific business issues, determining their costs, and ascertaining whether poor data quality is the root cause. After setting data SLAs to enhance data quality, evaluate the updated cost of the issue to the business. This quantification of the impact of bad data on various aspects of the business allows organizations to measure the tangible benefits of implementing data observability solutions.

The second step involves calculating the initial investment or cost associated with implementing data observability. This includes evaluating costs related to people, processes, and technology, considering both short- and long-term expenses in these categories to gain a comprehensive understanding of the total investment required. Discerning the exact costs can be intricate due to the multifarious elements involved in data systems, necessitating a holistic approach to assess both tangible and intangible costs accurately.

While quantifiable metrics are essential, it's also pivotal to consider less quantifiable metrics and the strategic value that data observability brings. This includes the ability to make informed business decisions, evaluate potential legal risks, grow retention, enhance organizational effectiveness, and foster a culture of transparency and accountability. These aspects, although not directly translatable to monetary values, provide significant insights into the intangible benefits and overall impact of data observability on an organization. In addition, it potentially improves the standing of the data team within the organization.

Moreover, the enhanced adaptability and competitive advantage gained through data observability can lead to increased market share and customer loyalty, which are invaluable in today's dynamic business environment. The cumulative effect of improved decision-making and operational efficiency can substantially impact the organization's profitability and long-term sustainability.

To measure the ROI of your first project effectively, define clear and measurable objectives and establish **key performance indicators (KPIs)** aligned with the project's goals. Conduct a baseline assessment to capture the current state of the identified KPIs and compare the post-implementation results to quantify the improvements achieved. Calculate the financial impact of these improvements and assess the total investment made in the project. Finally, calculate the ROI by comparing the financial benefits obtained with the total investment made, and document any qualitative benefits observed to provide a comprehensive view of the project's impact.

In conclusion, balancing the tangible and intangible benefits, along with a clear understanding of costs and strategic values, ensures a holistic assessment of the investment in data observability. This comprehensive approach aligns the assessment with the organization's overarching goals and values, providing a true reflection of the value derived from investing in a data observability solution.

Scaling with data observability

In this section, we will see how to scale with the implementation of data observability in a maximum of your pipelines. Once data observability is approved in the organization, the data team will work on including it in a maximum of projects. This has to be done procedurally to be sure to not fall into the pitfalls we listed previously. To do so, we will distinguish what to do with the new projects from what to do with legacy ones.

8. What should I do with new projects?

Once the incorporation of data observability is sanctioned within an organization, it necessitates being integrated as a pivotal phase in the formulation of every project.

To facilitate this, the validation phase of a project must encompass the incorporation of an apt data observability agent. Should a data discrepancy be identified in any of these projects, mechanisms should be in place to effortlessly trace the root cause. Additionally, if an agent is found to be absent, it should be straightforward to locate the accountable party to rectify this deficiency.

This procedure is not exclusive to new initiatives but should also be retroactively applied to existing projects undergoing modifications, such as those transitioning to a new library. The inclusion of a data observability referent in this process is crucial. This referent serves as a focal point, ensuring that data observability principles are consistently applied and upheld, thereby fostering an environment of accountability and continuous improvement in data management practices across the organization.

9. How should I deal with legacy projects?

Depending on the size of your organization, implementing data observability everywhere may be a tedious task. To do so, a task force needs to be created and needs to prioritize the job to be done.

A good way of doing this is to start monitoring the data with data quality. This can be done by applying at scale scanning of the data. Thanks to this, immediate visibility will be enabled, and SLI can be created.

After this, priorities should go to applications that create the data sources where issues are detected. In doing so, the task force will progressively create a map of its pipelines thanks to data observability being enabled. Nevertheless, for less documented pipelines, it may become difficult to introduce observability if the applications are not known. Finally, the task force needs to weigh up the pros and cons of putting an application under the data observability umbrella. A very old technology may be difficult to maintain while the team expects to migrate to a more recent one.

With these questions being answered, we can now create the data observability checklist. This tool, to be used to implement a data observability tool, will help you in the first phase of implementation, and at scale when data observability becomes part of your daily work.

Summary

This chapter delved into the intricate process of implementing and scaling data observability within organizations, emphasizing the common pitfalls faced during the integration of observability.

We have seen the main challenges, which are the control of costs, the overhead with other jobs, the security concerns, the increase in complexity of the architecture, the trade-off to be handled with legacy systems, and finally, the information overload that teams can experience. We have also seen that all these challenges can be overcome and the risks mitigated.

Then, we listed the questions a data team must answer during observability implementation. The list covered the criteria for selecting the appropriate project and observability tool, considering aspects such as security, compliance, cost, integration, data retention, intelligence, and customization. The discussion on costs explored various strategies, including open source solutions, in-house development, vendor solutions, and hybrid approaches, each with its unique financial implications.

The chapter provided insights into assessing the return on investment for data observability, emphasizing the need to consider tangible and intangible benefits, strategic values, and the overall impact on the organization. It stresses the importance of clear and measurable objectives, KPIs, and a baseline assessment to quantify improvements effectively.

The process of scaling data observability is explored, distinguishing between new and legacy projects. For new projects, the integration of data observability is crucial from the validation phase, with the inclusion of a data observability referent to ensure consistent application of data observability principles. For legacy projects, the chapter suggests starting with data quality monitoring and progressively creating a map of pipelines, weighing up the pros and cons of including each application under the data observability umbrella.

In the next chapter, we will focus on how we can implement a technical roadmap to develop data observability within an organization, offering you a more contextual and applied understanding of data observability in practical scenarios.

10
Pathway to Data Observability

Throughout this book, we have seen how data observability can be included in diverse industry standard tools. In the first part of this chapter, we will share a technical roadmap so that you can include data observability in your applications based on what we've observed from many past projects that we've conducted over the years.

This technical roadmap will also cover what can be done once pure observability has been implemented, as well as what can be added to complete the picture and make the most of it. We will discover the links with data management, AI, ML, and data quality programs.

The second part of this chapter will focus on applying what you have learned in this book by considering a company example and explaining how you can implement data observability in it.

This final chapter covers the following topics:

- Technical roadmap to include data observability
- Project

Technical roadmap to include data observability

In this section, we will describe how you can add step-by-step observability to your data applications while summarizing all the elements covered in this book. We will see what we need to consider when starting with data observability from a technical point of view. We will cover the following topics:

- Allocating the right resources to your data observability project
- Defining clear objectives with the team
- Implementing data observability in applications
- Continuously improving observability
- Scaling data observability

Let's get started!

Allocating the right resources to your data observability project

To implement data observability in your data applications, allocating the right resources is key to success. Such resources include budget, staffing, tools, and time.

Budget

First of all, defining a budget for data observability can be tricky. From our experience, we often see this budget as taken from the data quality budget. However, as we have seen in this book, augmented data quality is different from data observability. Both data observability and data quality are critical components of a successful data strategy, and compromising one for the other can have negative consequences.

The budget should contain the costs of the platform and tool you will use, the potential new hiring or costs of training, or consulting services.

Staffing

To implement a data observability project, the ideal team should contain the following elements, under the supervision of a project manager:

- **Data architects and data engineers**: These are the people who are in charge of creating the data pipeline, and are also responsible for its quality. They will implement data observability in the applications and design how the observability metrics will be retrieved. During the project, they will assume the majority of the workload, from collecting the logs to programmatically defining the rules.

- **Data/business analysts**: These people are often the customers of the data pipeline. They need to be consulted to set up the expectations and data quality rules the pipeline should respect.

- **Site reliability engineers**: These people are often at the frontline when a data issue happens, which means their inputs can't be neglected in this kind of project.

On top of this, training and evangelization must be foreseen in the company for it to foster a data observability culture.

Tools

To implement effective data observability, it's important to select the right tools that will suit your particular needs. Let's look at some key factors to consider when selecting tools for data observability.

First, the size and complexity of your data pipeline will have a direct impact on the type of tools you'll need for data observability. For example, if you're dealing with a large, complex data pipeline that covers multiple systems and technologies, you'll use tools that can handle large volumes of data, integrate with a variety of platforms, and support complex data transformations.

Your organization may be using a variety of technologies, such as cloud BI tools or frameworks. Your data observability tools should be able to support the technologies you're using, and ideally integrate easily with them. Also, as your organization grows and your data pipeline becomes more complex, your tools needs will evolve as well. You'll need tools that can scale with your organization and handle increasing amounts of data without sacrificing performance or reliability.

Finally, different data observability tools may use different techniques and resources to monitor and analyze data, such as log analysis or manual data review. You'll need to consider which techniques and resources are best suited to your particular use case and select tools that support those techniques.

Time

Data observability implementation is a continuous process, but its impact on the team is often low after the initial implementation.

Data observability is a step in the general process of pipeline creation, and if it's well standardized and formalized, it should not require a lot of time from your team. Once the framework has been standardized, the data (product) owner typically takes on the responsibility of understanding and detailing the requirements within that framework.

Even for a first implementation, depending on the tool you'd like to use, the first results can be obtained in less than a day of work.

Defining clear objectives with the team

Before starting the implementation, clear objectives need to be defined in the team. First, a pipeline has to be chosen. Second, the KPIs for the project have to be defined so that you can define the success criteria of the data observability project.

These objectives can have an impact on the resources that are allocated to the project.

Choosing a data pipeline

To choose the right pipeline, as stated in *Chapter 9*, you can start from a well-known pipeline or include data observability in current development. From the technical point of view, the choice of the pipeline will depend on the tool you selected and the integrations it has with the data sources and data applications of your pipeline.

Setting success criteria with the team and stakeholders

The success criteria are important to evaluate the outcome of the observability project. Knowing what needs to be done will help you decide on the right budget and resources.

Measurable success criteria

The focus should be on clear criteria that can be measured, for both the business and the technical team. For instance, if the team knows how many hours they spend solving data issues, it is important to notice how data observability could help in reducing this amount.

From a technical perspective, the focus should be on how fast the team can achieve 100% coverage of the pipeline, and what elements are made observable during the project.

From a technical point of view, success can be measured by knowing about the following:

- How many days of work are needed to implement data observability
- How many issues were discovered once the implementation was completed
- Whether all the necessary observations were retrieved
- Data sources
- Applications
- Lineages
- Metrics

The criteria for metrics need to be defined by the business. Inside these criteria, several levels of success can be defined. There is a tradeoff between the complexity of the data observability solution and its cost.

From what we experienced during the technical projects we conducted, complexity increases with the level of detail you want. *Figure 10.1* shows how complexity increases with events:

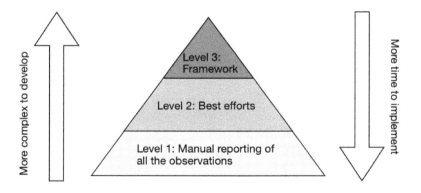

Figure 10.1: Level of technicality in data observability

Figure 10.1 shows the three levels of data observability we can achieve in application frameworks:

- The first level corresponds to manually logging the observations in the code when you encounter them. This requires a high level of attention from the developer so that they can ensure that all relevant observations are logged and not overlooked.

- The second level adds some abstraction and automation to the process. For instance, the developer could report the inputs and outputs of a data process, and the framework would do the rest by mapping the data sources and computing the metrics. This level of observability can be achieved through the use of libraries or frameworks that provide logging and monitoring capabilities, making it easier to collect and analyze data observations.

- In the third level, which is the most advanced, the highest level of automation is achieved. The framework detects all the data transformations at runtime with minimal configuration and without significant code changes.

The more you climb the levels of the pyramid, the higher the initial complexity to create the framework. However, a level 2 or 3 approach is less intrusive in the development life cycle at the expense of higher initial costs, whether it is to acquire a vendor solution or to build it internally. Lower levels mean more time is spent on the implementation within the data application itself. Nevertheless, a level 3 approach is advised when implementing data observability at scale as it requires few changes in the code and is easier to spread in the company's culture. Using a full observability framework, the developer can continue their work without caring much about technical data observability. Most of the vendors' solutions will target a level 3 approach.

That being said, the level of technical investment needed to succeed in the project is an important success criterion.

Implementing data observability in applications

Once the pipeline and tools have been decided on, you must start implementing them. We advise that you do this incrementally so that you can keep records of the time spent to activate each of these features, as well as the potential impact it has on the performance.

Step 1 – detecting the context, data sources, and data source lineages

The first observability element you should track is the context and the data sources used in the application. This is a prerequisite to ensure you can identify the location of the issue when it's detected and how to solve it. These elements can also be decomposed in smaller steps. For instance, if you agree that field-level lineage is not a key success criterion, you can delay the computation of this detailed lineage.

Step 2 – computing the metrics

Metrics are key to implementing a lot of data quality expectations. Once the lineage has been validated, you can add generic or custom metrics to the process. The more metrics you gather, the more you'll be covered in case of a data issue.

Step 3 – creating or validating the rules

This step consists of implementing data quality rules as per the business or suggested by the observability tool.

Continuously improving observability

Once your first pipeline is data observable, you can start thinking about processes to periodically review the collected metrics and rules.

First, on top of the observations you collect, if your data observability tool isn't doing it yet, make sure you regularly visualize your data. This will help you detect anomalies at a glance. For the rest, use the techniques that were discussed in *Chapter 3*.

Scaling data observability

Once the first pipeline has been validated, your team will want to scale the adoption of data observability across the company pipelines. This means you will need to integrate it in new as well as legacy pipelines.

For this, we advise that you follow these constraints:

1. For each new pipeline or application of a pipeline, an owner should be designated. This person is responsible for implementing data observability within a defined scope and must validate that it's delivered as a data product.

2. When an application is maintained, such as to migrate to a new tool or technology, the same method applies: a person is in charge of the data observability implementation.

3. For older applications and pipelines, which are not covered in the scope of categories 1 and 2, an application owner needs to be defined. The owner will be the one who's consulted when the service reliability engineer or data engineer suspects there's an issue with data coming from the application. They can use a catalog to list the following:

    -   The data sources used in the application

    -   The frequency of the run

    -   The code's location and the version of the application

The idea behind this process is that during root cause analysis, if the origin of the data is unknown, the application owner can answer any question and help with constituting the lineage.

We must stress the importance of the responsibility of deploying observability at scale. Once you've done this, the company culture should include its best practices so that if an issue is discovered in the new application and we can't find any linked lineage, the application owner can be questioned so that we know why there is a missing part.

Moreover, it's not because it's a legacy system that it can't be monitored. For all those applications, we advise using scanners to keep an eye on the inner metrics.

Once data observability has been implemented, you can start thinking about how the company can leverage the huge amount of information your observability tool contains.

Using observability for data catalogs

Data observability creates a dynamic database of all data usage across the company. Thanks to this, you can use observability to improve your data management processes.

A data catalog needs to be manually updated and may contain obsolete data source references. With direct reports from within the application, you can count on observability to provide an exhaustive view of the data sources that are used in your monitored applications. If a schema changes or a new data source is added, the element is directly updated in the data source catalog.

This helps significantly reduce the TCO of the data catalog while enhancing its completeness since filing and updating tasks can be automated.

Using observability to ensure ML and AI reliability

For a lot of ML libraries, the quality of the data is very important. A model can work with a certain distribution of the data and deliver poor results once in production as another distribution arises. Let's look at a simple example of an e-commerce model that's been trained on 2020 data; this can be influenced by the global lockdown and the changes in consumption behaviors. As the data goes back to normal, you can experience incorrect assumptions regarding the distribution that affect the robustness of the model.

That being said, note that for ML models, observability can be used at different levels to ensure the validity of a model over time.

First, observability should register the metrics on which the model has been trained. This means the metrics computed on the gold matrix, before the training phase, and the metrics based on the performance of the model, such as the root mean square error or the R2.

Then, when a model is run in production, the outcome of it, namely the prediction, should also be monitored and SLOs need to be attached.

In the end, the output of a model can be considered as a data source, with corresponding monitoring. Custom metrics can be used to compare the robustness of the model over time. This is done by comparing the model's performance in terms of training metrics and production metrics.

Using observability to complete a data quality management program

As stated in *Chapter 2*, data observability enhances data quality. Data quality focuses on detecting issues in used or produced data sources. Data observability allows you to not only reduce the time to detect but also fasten the resolution of the data issues.

Thanks to data observability, loads of data sources can be accessed within the frame of the running application. You do not require any additional security constraints as the application creating the metrics is also the one that's used to perform the business logic.

Now, let's see how this technical roadmap could be included in a project.

Implementing data observability in a project

In this section, we will learn how to implement data observability at scale in an organization, using the technical roadmap presented in the first part of this chapter.

Here, we will provide a macro view, at the organization level, of how observability will be progressively adopted. For the micro level, which includes inside a project, pipeline, or data product, please refer to *Part 2 – Implementing Data Observability*.

We will consider a company called PetCie, a pet store company that's active in five countries and produces and sells dog food to resellers.

This company currently has three projects:

- **Sales reporting**: A financial dashboard that's used by the CFO so that they can follow the day-to-day sales to the resellers. This report is the outcome of a Tableau dashboard that's linked to a PostgreSQL table and fed by a Spark application. They plan on changing Tableau into a Qlik process this quarter.

- **Production forecasts**: This model uses various internal and external data sources to forecast the weekly objectives of their five factories across the globe. They use an Amazon pipeline with S3 buckets to collect the daily production data, with a Lambda layer to a PostgreSQL database, and a GluonTS model to predict each week where production is needed. The model also uses seasonal indicators from external providers. This model has been running in production for more than 1 year.

- **Marketing qualification**: This pipeline creates a list of lead resellers that the sales team must contact and comes from various sources: partners and owned websites, registration forms, web scraping, and so on.

So, let's dig into the projects of this organization!

Resources and the first pipeline

At PetCie, the team is composed of a data engineer, who consults the CFO to set the business KPIs, and the engineering team lead, who sets the success criteria.

The team has decided that they want to start by implementing the sales reporting project since it will be coordinated with the Qlik migration. The team would like to gradually implement observability in each process when developing new ones or migrating to new tools.

They have decided to buy a solution that has a minimal impact on the other projects so that the team can focus on creating value with other projects.

Finally, they have decided to spend 4 days working on data observability in 1 month.

Success criteria for PetCie's implementation

The team has measured that, with the current implementation, the CFO detects an issue every 20 days on average. The first success criterion the team wants to achieve is to avoid garbage in/garbage out, so they want to ping the CFO ahead of any issues. Moreover, they want all of those issues to be explainable so that they can perform efficient root cause analysis.

From a technical point of view, the team needs to validate the implementation for the following aspects:

- The Qlik program, which is based on Postgres
- Spark Jobs

The team has decided that the most important aspect is to retrieve the data source that's used, its table lineage, and a subset of metrics to support the SLO that will be needed to build trust for the data consumer. This first phase should not take more than 10 man days.

The team wants the process to be repeatable and wishes to avoid adding complexity to the engineer's code. The structure of the company allows each project team to build their pipeline based on available datasets and frameworks. Adding observability to the framework is a good entry to later scale in the company.

The implementation phase at PetCie

After defining the SLO with the business team, the engineer starts the implementation. First of all, they retrieve the resources and the lineages from the Qlik application. However, the tool that was initially bought couldn't compute metrics on the progress table that's used as input. Therefore, the team also implemented an in-house scanner approach to scan the table. Thanks to this, they were able to fulfill the objectives set by the business team.

Right after this first implementation, the data engineer implemented observability in the Spark Jobs feeding the progress table.

Continuously improving observability at PetCie

Once all the applications were integrated, they began to see the first results. Errors were detected, and the CFO could be informed in time so that they never used bad quality data afterward.

The application's success could also be validated since every error, if it came from the Spark application, could be solved in less than 1 day.

The team decided to implement a process with the CFO to periodically review the status of the pipeline and define new objectives. This is done each time an issue arises but is not detected beforehand by the system.

Deploying observability at scale at PetCie

PetCie has decided to implement data observability in new or revamped applications. To keep an eye on other applications, they have implemented scanners on the PostgreSQL database to periodically check the quality of its tables.

Outcomes

In summarizing the fictional case of PetCie's approach to implementing data observability, we can draw several key insights. PetCie's phased and coordinated strategy, which starts with the sales reporting project, allowed for a focused and measured introduction to observability. By prioritizing minimal impacts on ongoing projects and committing dedicated time to observability, PetCie was able to integrate this new discipline without disrupting existing workflows.

The pragmatic approach of purchasing a solution to kickstart this process underscores the importance of leveraging external expertise to complement internal efforts. This decision was crucial in PetCie achieving their goal of preempting data issues and ensuring that any arising problems could be explained and addressed promptly.

The technical implementation, which is characterized by a mix of purchased tools and custom solutions, such as the scanner for the progress table, exemplifies the balance of adaptability and technical acumen that's required for such initiatives. PetCie's commitment to making the process repeatable and avoiding unnecessary complexity in engineers' code reflects their deep understanding of the need for sustainable and scalable solutions within the company's ecosystem.

The outcome of this initiative – a significant reduction in issue detection time, the ability to inform the CFO proactively, and a streamlined process for continuous review and improvement – demonstrates the value of a well-executed data observability strategy.

Summary

This chapter summarized all the steps a company must consider to deploy data observability in their pipelines. We have seen that the following technical roadmap can be set to ensure the success of the implementation:

1. **Allocate the right resources to your data observability project**: This consists of evaluating the right budget, staffing the right people, choosing the right tools, and defining a good period to start a first implementation.

2. **Define clear objectives with the team**: We saw how objectives can be set from a business and a technical point of view.

3. **Implement data observability in your applications**: During this process, we saw that success can be achieved by going incrementally in the implementation.

4. **Continuously improve observability**: Data observability is not a one-time process – it must be periodically reviewed to ensure permanent success.

5. **Scale data observability**: Finally, we saw how observability can be scaled in the company by proposing a tactic to gradually make all pipelines observable.

As we turn the final page of our collective exploration into the vibrant world of data observability, let's pause to reflect on the journey we've embarked upon together. From the foundational principles laid out in the opening chapters to the deep dives into implementation strategies, our path has been rich with learning and discovery.

As we've come to understand, data is not merely a static asset but a pulsating, living force within our organizations. It speaks volumes about the health and performance of our systems, offering insights that, when harnessed through data observability, can lead to remarkable achievements. Together, we've learned how to interpret the language of data, anticipate the needs of our systems, and act proactively to ensure its reliability, quality, and trustworthiness.

Our collective hope is that this book has not only expanded your understanding but also kindled a passion for the endless possibilities that data observability offers. Whether you are a data engineer, a business analyst, a data scientist, or someone who harbors a deep fascination regarding data systems, we trust that you've found valuable insights and inspiration within these pages.

Thank you for being an integral part of this adventure. May the knowledge and insights you've gleaned light your way forward and may your endeavors in data observability be fruitful and rewarding.

Index

Packtpub.com

Subscribe to our online digital library for full access to over 7,000 books and videos, as well as industry leading tools to help you plan your personal development and advance your career. For more information, please visit our website.

Why subscribe?

- Spend less time learning and more time coding with practical eBooks and Videos from over 4,000 industry professionals
- Improve your learning with Skill Plans built especially for you
- Get a free eBook or video every month
- Fully searchable for easy access to vital information
- Copy and paste, print, and bookmark content

Did you know that Packt offers eBook versions of every book published, with PDF and ePub files available? You can upgrade to the eBook version at packtpub.com and as a print book customer, you are entitled to a discount on the eBook copy. Get in touch with us at customercare@packtpub.com for more details.

At www.packtpub.com, you can also read a collection of free technical articles, sign up for a range of free newsletters, and receive exclusive discounts and offers on Packt books and eBooks.

Other Books You May Enjoy

If you enjoyed this book, you may be interested in these other books by Packt:

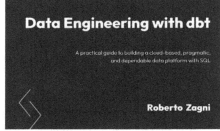

Data Engineering with dbt

Roberto Zagni

ISBN: 978-1-80324-628-4

- Create a dbt Cloud account and understand the ELT workflow
- Combine Snowflake and dbt for building modern data engineering pipelines
- Use SQL to transform raw data into usable data, and test its accuracy
- Write dbt macros and use Jinja to apply software engineering principles
- Test data and transformations to ensure reliability and data quality
- Build a lightweight pragmatic data platform using proven patterns
- Write easy-to-maintain idempotent code using dbt materialization

Azure Data Engineering Cookbook - Second Edition

Nagaraj Venkatesan, Ahmad Osama

ISBN: 978-1-80324-678-9

- Process data using Azure Databricks and Azure Synapse Analytics
- Perform data transformation using Azure Synapse data flows
- Perform common administrative tasks in Azure SQL Database
- Build effective Synapse SQL pools which can be consumed by Power BI
- Monitor Synapse SQL and Spark pools using Log Analytics
- Track data lineage using Microsoft Purview integration with pipelines

Packt is searching for authors like you

If you're interested in becoming an author for Packt, please visit authors.packtpub.com and apply today. We have worked with thousands of developers and tech professionals, just like you, to help them share their insight with the global tech community. You can make a general application, apply for a specific hot topic that we are recruiting an author for, or submit your own idea.

Share Your Thoughts

Now you've finished *Data Observability for Data Engineering*, we'd love to hear your thoughts! Scan the QR code below to go straight to the Amazon review page for this book and share your feedback or leave a review on the site that you purchased it from.

https://packt.link/r/1-804-61602-8

Your review is important to us and the tech community and will help us make sure we're delivering excellent quality content.

Download a free PDF copy of this book

Thanks for purchasing this book!

Do you like to read on the go but are unable to carry your print books everywhere? Is your eBook purchase not compatible with the device of your choice?

Don't worry, now with every Packt book you get a DRM-free PDF version of that book at no cost.

Read anywhere, any place, on any device. Search, copy, and paste code from your favorite technical books directly into your application.

The perks don't stop there, you can get exclusive access to discounts, newsletters, and great free content in your inbox daily

Follow these simple steps to get the benefits:

1. Scan the QR code or visit the link below:

https://packt.link/free-ebook/9781804616024

2. Submit your proof of purchase.
3. That's it! We'll send your free PDF and other benefits to your email directly.